# Mouse Mammary Tumor Virus

Edited by
P.K. Vogt and H. Koprowski

With 12 Figures

Springer-Verlag
Berlin Heidelberg NewYork Tokyo 1983

Professor Dr. PETER K. VOGT
University of Southern California
School of Medicine
Department of Microbiology
2025 Zonal Avenue HMR 401
Los Angeles, CA 90033, USA

Professor Dr. HILARY KOPROWSKI
The Wistar Institute
36th Street at Spruce
Philadelphia, PA 19104, USA

ISBN-13: 978-3-642-69359-5          e-ISBN-13: 978-3-642-69357-1
DOI: 10.1007/978-3-642-69357-1

© by Springer-Verlag Berlin Heidelberg 1983.
Library of Congress Catalog Card Number 15-12910
Softcover reprint of the Hardcover 1st edition 1983

Typesetting, printing and bookbinding:
Universitätsdruckerei H. Stürtz AG, Würzburg
2123/3130-543210

# Current Topics in Microbiology 106 and Immunology

Editors

M. Cooper, Birmingham/Alabama · P.H. Hofschneider,
Martinsried · H. Koprowski, Philadelphia · F. Melchers, Basel
R. Rott, Gießen · H.G. Schweiger, Ladenburg/Heidelberg
P.K. Vogt, Los Angeles · R. Zinkernagel, Zürich

# List of Contributors

COHEN, J.C., Departments of Medicine and Biochemistry, Louisiana State University Medical Center, New Orleans, LA 70112, USA

DICKSON, C., Imperial Cancer Research Fund Laboratories, Lincoln's Inn Fields, London, WC2A 3PX, U.K.

MICHALIDES, R., Department of Virology, Antoni van Leeuwenhoekhuis, The Netherlands Cancer Institute, Plesmanlaan 121, 1066 CX Amsterdam, The Netherlands

NUSSE, R., Department of Virology, Antoni van Leeuwenhoekhuis, The Netherlands Cancer Institute, Plesmanlaan 121, 1066 CX Amsterdam, The Netherlands

PETERS, G., Imperial Cancer Research Fund Laboratories, Lincoln's Inn Fields, London, WC2A 3PX, U.K.

RINGOLD, G.M., Department of Pharmacology, Stanford University, School of Medicine, Stanford, CA 94305, USA

TRAINA-DORGE, V., Department of Microbiology, Tulane University Medical Center, New Orleans, LA 70112, USA

VAN OOYEN, A., Department of Virology, Antoni van Leeuwenhoekhuis, The Netherlands Cancer Institute, Plesmanlaan 121, 1066 CX Amsterdam, The Netherlands

# Table of Contents

*Indexed in Current Contents*

# Proteins Encoded by Mouse Mammary Tumour Virus

CLIVE DICKSON and GORDON PETERS

# 1 Introduction

## 1.1 Aims and Perspective

It is the purpose of this article to review what is currently known about the proteins encoded by the genome of mouse mammary tumour virus (MMTV). Ideally such a review should encompass the size, physical properties, location, mode of expression, processing, and, where possible, the function of these proteins, but although this depth of knowledge is now becoming available for some other retroviruses whose complete nucleotide sequences

Imperial Cancer Research Fund Laboratories, Lincoln's Inn Fields, London, WC2A 3PX, United Kingdom

Current Topics in Microbiology and Immunology, Vol. 106
© Springer Verlag Berlin · Heidelberg 1983

have been determined, the picture for MMTV is still far from complete. Despite being the subject of detailed investigation for almost 50 years and historically representing the first isolation of a retrovirus from mammalian sources, MMTV remains among the less well understood of the retrovirus family. As will be discussed in more detail later, the reasons for this are largely practical, reflecting in some measure the complex and rather unusual biology of the system. However, it is these unusual features which have sustained the intense interest in the virus since its first isolation.

## 1.2  Distinctive Features of Mouse Mammary Tumour Virus

Clearly one of the principal motives for studying MMTV is its unique tissue tropism. Although there are reports of MMTV antigen production in a number of body tissues (see NANDI and McGRATH 1973; HILGERS and BENT-VELZEN 1978), the alveolar epithelial cells of the mammary gland are both the predominant site for replication and apparently the exclusive site for the tumorigenicity of the virus. It has been rigorously established that MMTV is the major causative agent in the high incidence of mammary carcinomas developed by some strains of mice (GROSS 1970; MOORE et al. 1979; CARDIFF and YOUNG 1980; MICHALIDES et al. 1981). Carcinomas are quite rare among retrovirus-induced tumours and apart from MMTV are more commonly associated with a few of the acutely oncogenic viruses carrying specific transforming genes. Although MMTV does have the potential for an additional gene not present in other retroviruses (see Sect. 5), it is unlikely that these sequences behave as an authentic oncogene. Both the progression of the disease, with an average latency of 6–9 months, and what we know from the molecular studies of MMTV indicate that it behaves as a non-acutely oncogenic, replication-competent retrovirus. However, the progression of the disease reveals a further unusual feature of MMTV, in that it can be markedly influenced by hormones. This is manifest at two levels, although it is possible that they are interrelated. Firstly, the development of tumours is influenced by the hormonal changes accompanying pregnancy (see GROSS 1970; MOORE et al. 1979; CARDIFF and YOUNG 1980), and, secondly, glucocorticoid hormones have been shown to modulate the expression of the MMTV genome at the level of RNA synthesis (see VARMUS et al. 1979). This latter feature is discussed in detail in another contribution in this volume.

The final, though historically the first, feature which sets MMTV apart from the majority of retroviruses is that it represents the prototype of a distinct morphological subclass, namely the B-type retroviruses (BERNHARD 1958; WEISS et al. 1982). The significance of this remains unclear, since despite these unusual features MMTV conforms to the basic pattern of genome structure and replication common to all retroviruses. As a result of this and the relative intractability of the MMTV system, much of our present understanding of the virus relies heavily on parallels and analogies drawn from other, more amenable retroviruses. To provide a framework for subsequent sections of this review and to establish principles which may not have been rigorously demonstrated for MMTV, we propose to summarize, briefly and simplistically, the characteristic features of all replication-competent retroviruses which are pertinent to this discussion. For more details, the reader is referred to WEISS et al. (1982).

## 1.3  General Features of Retroviruses

The genetic material of retroviruses comprises two apparently identical copies of single-stranded RNA which in mature virions occurs as a 60 to 80 S dimeric complex containing a number of associated cellular tRNAs. Within the internal core of the virus particle, this RNA is associated with several low-molecular-weight structural proteins and a few molecules of the virally coded, RNA-dependent DNA polymerase. This enzyme is responsible for the "reverse transcription" of viral RNA into DNA. Although the sizes of the internal proteins vary among different viruses and their precise arrangement within the virion is still unknown, most retroviruses contain four major structural proteins, plus one or two minor additional species in some instances. The viral core of ribonucleoprotein is enclosed within a unit membrane, derived from the host cell during budding. Associated with this membrane are two virally coded proteins, one or both of which are glycosylated, and which frequently appear as spikes or projections on the outer virion surface as visualized by electron microscopy. These envelope proteins are presumably involved both in virus assembly and budding and in penetration of the host cell membrane during infection.

Once inside the host cell, the genetic information in the viral RNA is copied into double-stranded DNA which becomes integrated into the chromosomal DNA of the host. This integrated form, called the provirus, is genetically colinear with the viral RNA, but is bounded by two long terminal repeat (LTR) segments created by the duplication of sequences from the 5′ and 3′ ends of the viral RNA (see Fig. 1). The integrated provirus forms the template for the synthesis by host enzymes of full-length viral RNA. This RNA is normally between 8000 and 10000 nucleotides long, polyadenylated, and in the positive sense; therefore, as well as forming the genetic complement of progeny virions, it can also serve as a messenger RNA for some of the viral proteins. The coding sequences in the RNA are organized into three distinct units, or genes. In a 5′ to 3′ order, these are *gag*, coding for all of the internal structural proteins (the name deriving from early studies on group-specific antigenicity); *pol*, encoding the RNA-dependent DNA polymerase; and *env*, for the viral components of the unit membrane envelope. All three genes are required for viral replication and deletion of or substitution of cellular information for any part of these sequences, as occurs in most acutely oncogenic retroviruses, renders that virus defective. The realization that the retroviral genome was organized

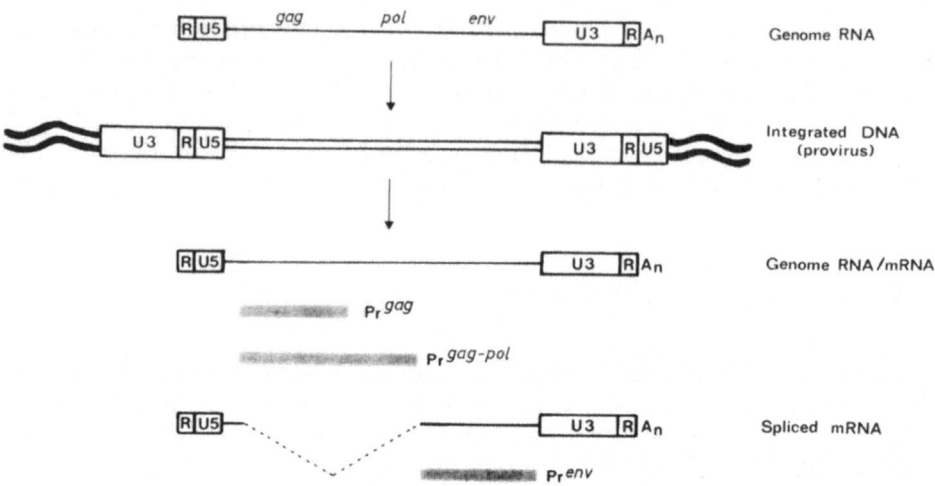

**Fig. 1.** Schematic representation of phases of the retrovirus life cycle. Genomic RNA is reverse transcribed into DNA to establish the provirus during the infectious phase. The integrated provirus then acts as a template for the transcription of new genomic RNA and mRNA, the major products of which are the polyprotein precursors as indicated (see *Weiss* et al. 1982 for further information)

into three such units was based on two types of observation: firstly, the genetic recombination pattern of conditional and non-conditional mutants of avian leukosis and sarcoma viruses, and, secondly, the recognition of discrete, polyprotein precursors as the primary translation products of each unit, as shown in Fig. 1. The organization, processing, and the constituent proteins which make up these precursors in the MMTV system form the subject matter for this review.

## 1.4 Methods of Study

As alluded to earlier, the MMTV system has proved to be rather intractable as compared with other retroviruses. In this section we intend to reflect on some of the reasons for this as part of a general discussion of the strategies so far employed. Probably the major obstacle to MMTV research has been, and still is, the lack of an appropriate in vitro assay for infectivity. Until relatively recently, the only source of the virus was either primary tumour tissue or milk from viremic mice, with the consequential need to maintain substantial breeding colonies of suitable strains of mice. Although the virus obtained in this way is infectious in vivo and capable of inducing mammary tumours with an incidence in excess of 90%, it appears to have an extremely low infectivity in tissue culture (LASFARGUES et al. 1974, 1976; VAIDYA et al. 1976). Even when successfully introduced into cultured cells, the amount of progeny virus released is often barely detectable and only

**Table 1.** Proteins of the MMTV virion

| Protein designation | Molecular mass range (kilo daltons)[a] | Percentage of total virion protein (range)[a,b] | Number ratio[c] | Precursor[gene] |
|---|---|---|---|---|
| gp52 | 47–55 | 26 (20–31) | 1 | Pr73[env] |
| gp36 | 33–38 | 18 (16–23) | 1 | Pr73[env] |
| p30 | 29–30 | 4 (2–7) | 0.26 | Pr110[gag] |
| p27 | 24–28 | 18 (16–20) | 1.3 | Pr77[gag] |
| pp21 | 17–23 | 4 (1–8) | 0.38 | Pr77[gag] |
| p14 | 13–18 | 8 (7–11) | 1.14 | Pr77[gag] |
| p10 | 9–12 | 9 (7–16) | 1.8 | Pr77[gag] |
| p100 | – | [d] | – | Pr160[gag/pol] |

[a] These figures were compiled from DICKSON and SKEHEL 1974; TERAMOTO et al. 1974; SARKAR and DION 1975; SARKAR et al. 1976; YAGI and COMPANS 1977
[b] Average of estimates determined using both protein staining and incorporation of radioactive amino acids. The total percentage composition is less than 100% due to additional proteins present in virion preparations but thought not to be encoded by the virus
[c] The number ratio relative to gp52 was calculated by dividing the percentage contribution of each protein species in the virus by its average molecular weight
[d] Although the reverse transcriptase probably represents a very small proportion of the total virion by weight, it is readily detected by enzyme assay

occasionally sufficient for biochemical studies (HOWARD et al. 1977). Moreover, the infected recipient cells display no obvious morphological alterations (in phenotype) and this, coupled with the hazards and duration of in vivo assays, has precluded any genetic studies based on the isolation of mutants.

The poor infectivity of MMTV in cultured cells may in part reflect our inability to propagate an appropriate target cell. Nevertheless, it has proved possible to culture and establish productive cell lines from virally induced mammary tumours (SYKES et al. 1968; LASFARGUES et al. 1972; OWENS and HACKETT 1972; RINGOLD et al. 1975). This, coupled with the discovery that glucocorticoid hormones will stimulate virus production by these cells (see RINGOLD, this volume), has provided conditions to secure adequate quantities of virus for a detailed structural analysis. Most investigators chose initially to estimate the number, sizes, and approximate quantities of proteins in mature virus (NOWINSKI et al. 1971; DICKSON and SKEHEL 1974; TERAMOTO et al. 1974; SARKAR and DION 1975; KIMBALL et al. 1976; YAGI and COMPANS 1977), and subviral particles (FELDMAN et al. 1973; CARDIFF et al. 1974; TERAMOTO et al. 1977a) using polyacrylamide gel electrophoresis. The consensus from these studies was that MMTV contains at least seven routinely identifiable proteins, the sizes and amounts of which are summarized in Table 1. In addition to these structural proteins, the Table includes the viral reverse transcriptase enzyme, the existence of which was inferred from assays of RNA-dependent DNA synthesis, rather than from acrylamide gel analyses. The diversity of conditions used for the analysis of these proteins, as well as the sources and purity of the virus stocks, led to considerable variation in their calculated molecular weights. Consequently, the protein designation in Table 1 represents a standardized nomenclature reflecting a consensus, and not necessarily actual molecular weights, in an effort to reduce the confusion in comparing results from different laboratories. However, the adoption of a standard nomenclature should not be taken to imply that the proteins under study in different laboratories are necessarily identical. Both the internal proteins and the glycoproteins have been shown to carry type-specific as well as group-specific antigenic determinants which have formed the basis of a number of radioimmunoassays for the different strains of MMTV present in various inbred mouse lines.

The availability of specific antisera and cell lines which produce MMTV have in combination provided the bulk of our current knowledge of the structure and expression of the MMTV proteins. Thus, it became possible to label cultured cells with radioactive amino acids or sugars and to prepare immunoprecipitates of the extracted cellular proteins. In this way sufficient quantities of the intracellular precursors and processing intermediates were obtained to perform detailed characterization by acrylamide gel electrophoresis and subsequent tryptic peptide mapping. The judicious use of inhibitors provided additional information, for instance, regarding the order of proteins on the precursors and the number of carbohydrate addition sites on the glycoproteins. The further application of cell-free protein synthesis systems to examine the coding potential of the viral RNA permitted both the ordering of the three known viral genes and the identification of an additional region of open reading frame (*orf*) which may represent all or part of a novel gene (discussed in detail in Sect. 5). As a result of these studies, a reasonably comprehensive picture of the organization of the MMTV genome and the proteins which it may encode has been established and will be presented below. However, a fuller understanding awaits the derivation of a complete nucleotide sequence for at least one of the infectious strains of MMTV. This requires the isolation of appropriate recombinant DNA clones. Progress in this direction has again been thwarted in the MMTV system by the difficulties, reported by several laboratories, in obtaining clones containing a particular region near the 5' end of the genome. Thus, our current sequence information pertains only to the *env* gene and the *orf* sequences in the LTR. These will be presented as parts of the following sections in which we deal in turn with the precursors, processing, and mature protein products from each of the four regions of the MMTV genome; *gag, pol, env,* and *orf.*

## 2  Mouse Mammary Tumour Virus *gag* Gene Products

### 2.1  Subviral Particles and Morphogenesis

Before discussing the individual *gag* gene products in detail, it is perhaps pertinent to review some of the properties of the naturally occurring and artificially produced subviral particles of which these proteins are major constituents. The earliest attempts to subclassify the retroviruses, based on their appearance in the electron microscope, placed MMTV into a distinct morphological group called B-type particles (BERNHARD 1958). Not only do the mature B-type viruses differ considerably in appearance from the more common C-type viruses, but their morphogenesis involves three recognizably different types of particle (DALTON et al. 1966; CALAFAT and HAGEMAN 1968). In thin-section electron micrographs of MMTV-producing cells, structures known as intracytoplasmic A-type particles are readily detected in the cytoplasm (BERNHARD 1958). These appear as two concentric rings of electron-dense material approximately 70 nm in diameter and have been shown to be immunologically related to at least some of the *gag* gene proteins (TANAKA et al. 1972; SMITH and WIVEL 1973; SARKAR and DION 1975; SMITH and LEE 1975; SARKAR and WHITTINGTON 1977; TANAKA 1977). At the plasma membrane (and often at other internal vesicular membranes), these A-type particles appear to adhere to the inner surface and probably instigate the budding process, as a consequence of which the particles acquire a unit membrane envelope. Very rapidly after budding, these extracellular immature particles change in morphology so that the core contracts to an electron-dense nucleoid of about 60 nm in diameter. In the final, mature virion this nucleoid is eccentrically located and is immediately surrounded by a fine "membranelike" structure which, in turn, is enclosed within the outer membrane envelope, giving a final particle diameter of about 110 nm. Conversely, it is experimentally possible to release the internal core from the virion by treatment with non-ionic detergents to remove the outer membrane. These artificially produced subviral particles, or "cores", have a higher buoyant density than intact virions and contain genomic RNA, reverse transcriptase, and a subset of the *gag* gene proteins (FELDMAN et al. 1973; TERAMOTO et al. 1977a).

## 2.2 Properties of the *gag* Proteins

One of the most abundant polypeptides in both virus and core preparations is p27; for this reason, p27 is presumed to be the major structural component of the virus core (DICKSON and SKEHEL 1974; TERAMOTO et al. 1974, 1977a; SARKAR and DION 1975; KIMBALL et al. 1976; YAGI and COMPANS 1977) (Tables 1, 2). Its abundance, coupled with its strongly immunogenic properties, has been exploited as the basis for several radioimmunoassays used for the detection of MMTV (PARKS et al. 1974; HENDRICK et al. 1978; TERAMOTO and SCHLOM 1978). Although a high proportion of p27 molecules appear to be phosphorylated (NUSSE et al. 1978; SARKAR et al. 1978), a property reflected in the heterogeneity observed in isoelectric point (pH 6.5 to pH 6.8) (NUSSE et al. 1980), the protein also contains a hydrophobic domain (MARCUS et al. 1978), consistent with its postulated role as the shell of the viral nucleoid.

**Table 2.** Properties of the MMTV *gag* gene proteins

| Protein | Properties | IEP[a] | Postulated functions |
|---------|-----------|--------|---------------------|
| p30 | Related to p14 | – | Unknown |
| p27/pp27 | Predominantly phosphorylated; contains hydrophobic domain | 6.5–6.8 | Major structural component of the core |
| pp21 | Major virion phosphoprotein; at least five levels of phosphorylation | 4.5–5.0 | Suggested regulatory and structural roles; none defined |
| p14 | Basic protein; binds to single-stranded DNA | 8.8 | Complexes with RNA; possible packaging function |
| p10 | Hydrophobic protein; contains palmitic acid | 7.5 | Associates with envelope; involvement in core assembly |
| p8 | Basic protein | – | Unknown |

[a] The isoelectric points (IEP) were taken from Nusse et al. (1980)

Although p27 is phosphorylated, the major virion phosphoprotein is pp21 (Nusse et al. 1978; Sarkar et al. 1978; Dion et al. 1979a). In isoelectric focusing, this protein behaves as a heterogeneous collection of about five distinct species, presumably resulting from variability in the degree of phosphorylation (Table 2; Nusse et al. 1980). It also has the capacity to exist in multimeric forms (Dion et al. 1979b), which has led to some suggestions that it contributes to the inner-membranelike structure between the core and the envelope. This would be consistent with both the location of phosphoproteins in other retroviruses and its apparent absence from viral core preparations (Teramoto et al. 1977a). However, pp21 is both hydrophilic (Marcus et al. 1978) and acidic (Nusse et al. 1980), so that it could be readily lost during the extraction procedures used to prepare cores and would seem unsuitable as a classical membrane protein. Moreover, in analyses of total virion proteins, it appears to be present in less than stoichiometric amounts, relative to other structural proteins (Table 1). Small amounts of the major phosphoproteins of avian and murine C-type retroviruses (pp19 and pp12, respectively) are found to specifically associate with their respective genomic RNAs (Sen and Todaro 1977; Sen et al. 1977), so that it remains possible that an analogous, but as yet undefined, regulatory function may exist for pp21 in MMTV virions.

A major component of core preparations is the highly basic protein, p14 (Teramoto et al. 1977a), which has the capacity to bind to single-stranded DNA (Arthur et al. 1978b) and confers this ability on the polyprotein precursor from which it is processed (Massey and Schochetman 1979). Analogous highly basic, low-molecular-weight proteins are also present in avian and murine C-type retroviruses, apparently bound to the genomic RNA to the extent that they can be isolated in the form of a ribonucleoprotein complex (Fleissner and Tress 1973). Thus, it is not unreasonable to assume that in MMTV p14 performs a similar function and may be instrumental in the packaging of viral RNA into the core (Table 2).

The most hydrophobic of the *gag* gene products is p10 (MARCUS et al. 1978) which, although it comprises up to 10% of the total virion protein (Table 1), is only poorly represented in core preparations (TERAMOTO et al. 1977a). Its absence from cores and two additional pieces of indirect evidence suggest that it may be closely associated with the internal side of the viral envelope (CARDIFF et al. 1978). First, MMTV-infected cells, made semipermeable with EDTA, were found to undergo lysis when exposed to anti-p10 serum and complement (MASSEY and SCHOCHETMAN 1979); second, both p10 and the putative transmembrane glycoprotein, gp36, can be labelled with radioactive palmitic acid, a fatty acid commonly attached to membrane proteins (R. NUSSE, personal communication). These properties and the location of p10 at the amino terminus of the major *gag* precursor (see below) have led to the suggestion that p10 may function in interactions between the internal protein precursor and the plasma membrane, thus facilitating core assembly and subsequent budding (Table 2; CARDIFF et al. 1978; DICKSON and ATTERWILL 1979).

In addition to the four major internal structural proteins described so far, some laboratories have reported the presence of two minor, presumably *gag*-related, components. The first of these, p30, has been clearly shown to be related to the basic protein, p14, since it contains all of the tryptic peptides characteristic of p14 (GAUTSCH et al. 1978; DICKSON and ATTERWILL 1979). While it is conceivable that p30 may simply represent a minor processing intermediate, the observation that the additional tryptic peptides of p30 which are absent from p14 are present in a minor, extended form of the *gag* precursor (see below) suggests that p30 is a functionally distinct moiety (DICKSON and ATTERWILL 1979). The status of the final protein, p8, remains uncertain and will possibly only be resolved when the complete nucleotide sequence of the MMTV *gag* gene is determined. It is deficient in methionine and has, therefore, gone undetected in studies utilizing radio-labelled methionine as a tracer. Moreover, its molecular weight remains open to doubt, since it migrates very close to the buffer front even in higher percentage acrylamide gels and is difficult to resolve from the more abundant p10 protein. Contamination of p10 by p8 might in part explain the apparent twofold over-abundance of p10 relative to other viral proteins (Table 1). Little can be said regarding either its function or location within virions, except that it is likely to be very basic in view of numbers of arginine and lysine residues detected by tryptic peptide mapping (DICKSON and ATTERWILL 1979).

## 2.3 Intracellular Precursors of the *gag* Proteins

As mentioned previously, the coding units of the retrovirus genome are expressed initially as polyprotein precursors. Thus, immunoprecipitation of

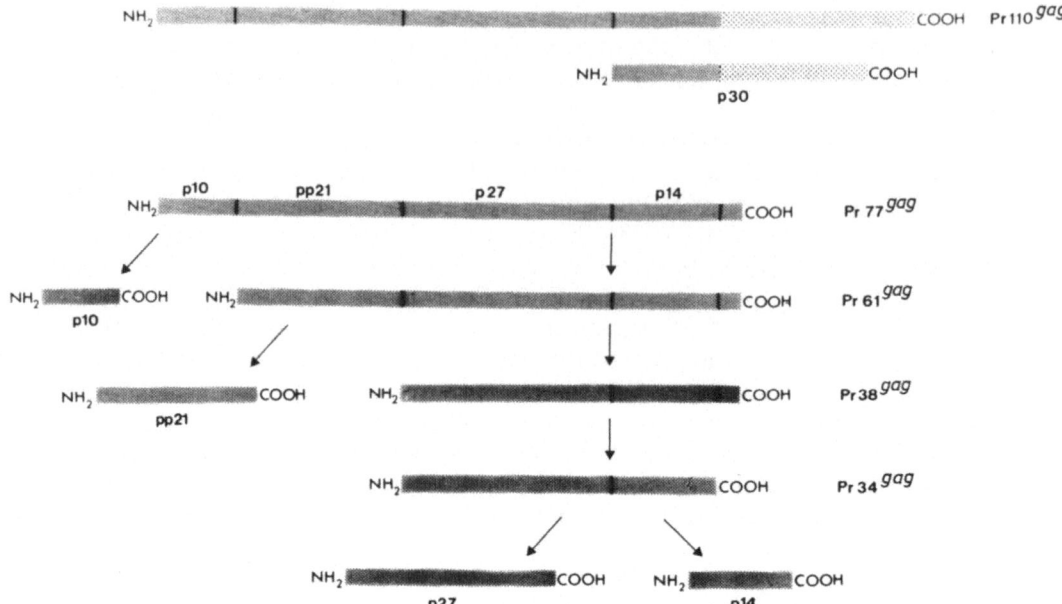

**Fig. 2.** Organization of the MMTV *gag* gene. The schematic diagram depicts the *gag*-related precursors and presumed processing intermediates detected in MMTV-infected cells. The composition and order of the individual viral proteins within each polypeptide species was deduced by pactamycin inhibition and tryptic peptide mapping. This permitted construction of the tentative cleavage pathway shown. The minor protein, p30, is presumed to originate from the extended form of the *gag* precursor, $Pr110^{gag}$. The p8 protein is not included in this figure in view of the uncertainty over its position and status as a viral protein

labelled extracts from MMTV-producing cells with, for example, anti-p27 serum led to the detection of a series of *gag*-related polypeptides ranging in size from 34 000 to 160 000 daltons (DICKSON and ATTERWILL 1978, 1979; NUSSE et al. 1978; RACEVSKIS and SARKAR 1978; SCHOCHETMAN et al. 1978; ANDERSON et al. 1979). However, after pulsing infected cells with either [$^{35}$S]-methionine or [$^{14}$C]-arginine and lysine, by far the major polypeptide detected using anti-*gag* sera is a 77000-dalton species, $Pr77^{gag}$. On SDS-acrylamide gels, this precursor actually migrates as a doublet, apparently a consequence of phosphorylation of a proportion of the molecules (NUSSE et al. 1978; RACEVSKIS and SARKAR 1979). The phosphorylation may be an obligatory step in the post-translational processing of the precursor, since the phosphopeptides detectable in $Pr77^{gag}$ are represented in the mature viral phosphoproteins. The fact that both forms of the doublet contain all four of the major structural proteins (p27, pp21, p14, and p10) and that a protein of this size and composition is the predominant product of in vitro protein synthesis programmed by MMTV virion RNA (see below) argues strongly for $Pr77^{gag}$ being the major *gag* precursor (Fig. 2). The additional lower-molecular-weight species detected intracellularly (Pr61, Pr38, and Pr34) are therefore presumed to be intermediates in the cleavage events required to generate the mature proteins (DICKSON and ATTERWILL 1978; RACEVSKIS and SARKAR 1978; MASSEY and SCHOCHETMAN 1979).

The use of monospecific antisera to the four major *gag* proteins and, more significantly, the detailed peptide mapping of the various processing intermediates combined to provide a reasonably comprehensive picture of the organization of the *gag* gene and permitted the derivation of a tentative cleavage pathway (DICKSON and ATTERWILL 1979) shown in Fig. 2. A further bonus of such studies was that they predicted the order of the proteins in the precursor as $NH_2$–p10–pp21–p27–p14–COOH (DICKSON and ATTERWILL 1979; MASSEY and SCHOCHETMAN 1979). This orientation was based partly on analogies with other retroviruses, placing the most hydrophobic protein at the amino terminus and the most basic at the carboxy terminus, but was later confirmed by pactamycin mapping studies (DICKSON and ATTERWILL 1980; SEN et al. 1980b). Pactamycin specifically blocks the initiation of protein synthesis, so that addition of this inhibitor during a pulse-chase experiment will effectively decrease the level of radioactivity incorporated into the amino terminal region of any polypeptide, relative to its carboxy terminus. In such experiments, the putative *gag* gene component, p8, appears to map on one or other side of the phosphoprotein, pp21 (not shown).

An interesting observation from pulse-chase experiments is that very little of the mature viral proteins can be detected in MMTV infected cells. After extended chase periods, the predominant intracellular intermediate – and the only one to accumulate during the chase – is a 34000-dalton species ($Pr34^{gag}$), which contains all of the tryptic peptides of the two most prominent core proteins, p27 and p14 (DICKSON and ATTERWILL 1978). Moreover, when virus particles are harvested from labelled cell cultures at short intervals (5 min) and if measures are taken to inhibit proteolysis, the major virion protein is $Pr34^{gag}$. Subsequent incubation of such virus preparations at 37 °C results in a decrease in the level of $Pr34^{gag}$ and a reciprocal increase in p27 (DICKSON and ATTERWILL 1978). This finding, and the virtual absence of p27 from infected cells, suggests that virions may bud with $Pr34^{gag}$ as their major core component and that proteolytic cleavage of this intermediate, to release p27 and p14, may be the biochemical basis for the post-budding maturation of the MMTV core. A thiol protease, which could be responsible for this and other steps in the processing of $Pr77^{gag}$, has been detected in MMTV virion preparations (SEN et al. 1980a). The detailed properties of this protease and whether it corresponds to one of the already established viral proteins remain to be determined, but it is clear that it demonstrates a marked specificity for the MMTV *gag* precursor, as opposed to those from other retroviruses.

As well as $Pr77^{gag}$ and the processing intermediates discussed above, MMTV-infected cell extracts contain two other *gag*-related polyproteins with molecular masses of 160000 and 110000 daltons. Both appear to include all of the tryptic peptides present in $Pr77^{gag}$ (plus additional characteristic peptides) (ANDERSON et al. 1979; DICKSON and ATTERWILL 1979) but from pulse-chase experiments and other considerations (see below) do not appear to represent precursors for Pr77$^{gag}$. By analogy with other retroviruses, the 160000-dalton species, Pr160$^{gag-pol}$, is presumed to represent the precursor for the RNA-dependent DNA polymerase and will be further discussed in the next section. In contrast, no analogies can be drawn for the Pr110$^{gag}$ species, since other retroviruses have no equivalent product. Pr110$^{gag}$ is not glycosylated and therefore does not correspond to the additional form of the *gag* precursor observed with the murine C-type viruses (see WEISS et al. 1982). Curiously, a number of the additional peptides of Pr110$^{gag}$, which are not present in Pr77$^{gag}$, appear to correspond to the characteristic peptides of the minor virion protein, p30, which are not shared with p14 (DICKSON and ATTERWILL 1979). Since p14 has been mapped to the carboxy terminus of Pr77$^{gag}$ and constitutes a subfraction of p30, it seems likely that Pr110$^{gag}$ represents an extended form of the major precursor generated by read-through of the termination codon at the end of Pr77$^{gag}$ (Fig. 2). The mechanism for such an event remains obscure but could conceivably involve either suppression or frame shifting by a specific tRNA or removal of the termination codon by RNA splicing. Unfortunately, the resolution of this issue will probably have to await the derivation of the nucleotide sequence in the region of the genome which has proved refractory to cloning.


## 2.4  In Vitro Expression of the *gag* Gene

An added complication in deciding the status of the three gag-related precursors is that all three can be synthesized in cell-free translation systems primed by MMTV-virion RNA. The relative proportions of Pr77$^{gag}$, Pr110$^{gag}$, and Pr160$^{gag-pol}$ observed in vitro mimic those seen in infected cells, with Pr77$^{gag}$ the predominant product (DAHL and DICKSON 1979; SEN et al. 1979). Moreover, kinetic experiments clearly indicate that synthesis of authentic Pr77$^{gag}$ is complete before Pr110$^{gag}$ and Pr160$^{gag-pol}$ are detectable, verifying that the two latter products are unlikely to be precursors for the major translation product, Pr77$^{gag}$ (DICKSON and PETERS 1981; SEN et al. 1981). In contrast to eukaryotic cellular mRNAs, the retrovirus genome, like those of some other RNA viruses, is polycistronic. However, unless breaks are introduced in the RNA, protein synthesis is generally restricted to the 5′ proximal gene, initiating at a single specific methionine codon adjacent to the preferred ribosome binding site. Thus, although it is conceivable that intact genome RNA prepared from MMTV virions does not represent a single RNA species, the detection of all three *gag*-related precursors in vitro, all containing peptides characteristic of the amino terminal protein, p10, suggests that they are independent primary translation products initiated at a single common site. While these in vitro studies shed no further light on the possible read-through mechanisms required to generate Pr110$^{gag}$ and Pr160$^{gag-pol}$, perhaps their major contribution has been to confirm that, as in other viruses, the MMTV *gag* gene is located at the 5′ end of the genome, presumably followed by *pol*. But even this conclusion must be qualified by the realization that the existence of Pr110$^{gag}$ casts some doubt on the precise delineation between *gag* and *pol*.


# 3 *pol* Gene Products


By definition, all replication-competent retroviruses encode an RNA-dependent DNA polymerase enzyme which is responsible for the reverse transcription of viral RNA into DNA. Although present in the virus particle, the polymerase is not a major structural component, the most precise estimate putting the number of molecules per virus particle at about 70 (PANET et al. 1975). As a result, the presence of the enzyme is more readily demonstrated by its activity than with detection by polyacrylamide gel electrophoresis. In common with other retrovirus polymerases, the activity of the MMTV enzyme is dependent on the presence of a divalent cation, a sulfhydryl reagent, deoxynucleotide triphosphates, and a template:primer complex. Although the natural primer for the MMTV reverse transcriptase is a cellular tRNA$^{Lys}$ (PETERS and GLOVER 1980), base-paired to the viral genome RNA near its 5′ end, a variety of synthetic polynucleotide:oligonucleotide combinations have been shown to serve as template:primers for in vitro assays. An important finding of these assays was that with most templates (the notable exception being poly rCm:oligodG) the MMTV enzyme displays a clear preference for magnesium over manganese as the optimal divalent cation (DICKSON 1973; HOWK et al. 1973; DION et al. 1974a; MARCUS et al. 1976). This is in marked contrast with the C-type murine retroviruses, whose polymerases show the inverse preference, and has, therefore, been exploited as an assay for the detection of MMTV against a background of C-type viruses (DION et al. 1974a).

By following enzyme activity, the MMTV DNA polymerase has been partially purified by a number of groups. In glycerol gradients, it sediments as a protein with a molecular mass of around 100000 daltons, but some discrepancy still exists regarding its polypeptide composition. DION et al. (1974b) described a single polypeptide chain of approximately 100000 daltons, whereas MARCUS et al. (1976) reported two subunits of 85000 and 50000 daltons, respectively, in general agreement with the two-subunit species isolated from intracytoplasmic A-type particles by KOHNO and ISHIHAMA (1979). Since the polymerases of other retroviruses have both one- and two-subunit structures, the nature of the MMTV enzyme cannot be inferred by analogy.

The lack of enough purified material to characterize has also precluded the raising of antibodies against the MMTV DNA polymerase. Consequently, little definitive work has been possible regarding the intracellular precursor for the enzyme. In the more amenable retrovirus systems it is clear that the *pol* gene is expressed as a high-molecular-weight fusion protein, containing virtually all of the characteristic tryptic peptides of both *gag* and *pol*. The most likely candidate in the MMTV system is $Pr160^{gag-pol}$, since it appears to meet most of the criteria of other analogous precursors. For example, $Pr160^{gag-pol}$ can be precipitated by anti-*gag* sera, it contains all of the tryptic peptides present in $Pr77^{gag}$ and $Pr110^{gag}$, it can be synthesized in vitro from intact genomic RNA, and the kinetics of synthesis are consistent with a read-through from *gag*. However, $Pr160^{gag-pol}$ contains only one or two methionine-labelled peptides not present in the other *gag*-related precursors, and no equivalent data is available for the mature polymerase. The absence of both immunological and peptide mapping data means that any conclusions regarding $Pr160^{gag-pol}$ must remain equivocal. Similarly, the mechanism by which read-through of the presumed termination codons at the ends of both $Pr77^{gag}$ and $Pr110^{gag}$ is achieved remains open to debate. Uncertainty also surrounds the processing required to convert $Pr160^{gag-pol}$ into the active enzyme. A potential 130000-dalton intermediate has been reported in cell extracts (MASSEY and SCHOCHETMAN 1979) but, by analogy with the murine C-type virions, it would seem likely that the final processing would be delayed until after budding to avoid releasing active reverse transcriptase in the cytoplasm of the already infected cell. Clearly our knowledge of the MMTV DNA polymerase is sketchy and littered with assumptions but there appear to be no major obstacles to future developments. These might include derivation of the nucleotide sequence upstream from the known sequence of *env* (see next section) and the subsequent possibility of raising antisera to synthetic peptides predicted from the sequence.

# 4 *env* Gene Products

## 4.1 Properties of the Envelope Glycoproteins

In common with other retroviruses, MMTV buds from the surface membrane of the host cell without causing lysis and in so doing acquires a unit membrane envelope. As with other types of enveloped virus, the segment of membrane surrounding the particle also contains two virally coded glycoproteins – in this case gp52 and gp36 (Table 1) – which together constitute approximately 40% of the total virion proteins (WITTE et al. 1973; DICKSON and SKEHEL 1974; PARKS et al. 1974; TERAMOTO et al. 1974; SARKAR and DION 1975). We now have a considerable amount of information about these two proteins, including their complete amino acid sequences, but our understanding of how they function in penetration of the host cell membrane and virus assembly is still in its infancy.

Several lines of evidence indicate that at least portions of both gp52 and gp36 are exposed on the outer surface of virus particles and probably comprise the projections or spikes detected with thin-section electron microscopy on mature virions and on the cell surface at the site of budding (TANAKA and MOORE 1967). For example, when the spikes are digested from the surface of the virion using proteases, the resultant "bald" particles appear to be deficient in gp52 and gp36, whilst retaining the normal complement of internal structural proteins (CARDIFF et al. 1974). Similarly, gp52 can be specifically stripped from the viral envelope by treatment at low pH, again resulting in bald particles (SARKAR et al. 1976). Radioactive labelling of tyrosine residues by lactoperoxidase or chloramine-T-catalysed iodination of intact virions also specifically tags gp52 and not gp36 (CARDIFF 1973; WITTE et al. 1973; PARKS et al. 1974), but this may in part reflect the numbers as well as accessibility of tyrosine residues in each of the glycoproteins. In contrast, labelling of carbohydrate side chains by galactose oxidation and reduction in the presence of [$^3$H]-borohydride indicates that at least parts of both proteins must be exposed on the virion surface (SHEFFIELD and DALY 1976; DICKSON and ATTERWILL 1980). Experiments aimed at determining the vicinal relationships between the virion components employing bifunctional cross-linking reagents readily detect homodimers and heterodimers of the two glycoproteins (DION et al. 1979b; RACEVSKIS and SARKAR 1980). The tentative conclusions from these various observations would be that gp36 acts as a transmembrane protein, anchoring the viral glycoprotein complex into the lipid bilayer; that gp52 is completely external but tightly associated with gp36; and that these two glycoproteins may form larger, oligomeric structures representing the spikes observed by electron microscopy. What we know of the properties of the two glycoproteins and their predicted amino acid sequences lends support to such conclusions.

In line with its supposed role as a transmembrane protein, gp36 is strongly hydrophobic, as determined physically (MARCUS et al. 1978) and as deduced from its (predicted) amino acid sequence (REDMOND and DICKSON 1983), which reveals two major regions of hydrophobicity at the amino and carboxy termini respectively (see Fig. 4). It is also glycosylated, comprising between 11% and 19% carbohydrate by weight (YAGI et al. 1978). This variability in sugar content presumably accounts for its heterogeneous electrophoretic mobility; in SDS-acrylamide gels it usually resolves into two major species with apparent molecular masses of 33000 and 37000 daltons. Both have similar methionine-labelled tryptic peptides (DICKSON et al. 1976) and, curiously, it is the faster migrating form which is the more heavily glycosylated (YAGI et al. 1978). Heterogeneity in both glycoproteins is also apparent in isoelectric focusing, where separation depends on charge, and up to five or six identifiable species of each glycoprotein can be resolved (NUSSE et al. 1980). Interestingly, the pI range for gp52, which contains only 9% carbohydrate by weight (YAGI et al. 1978), moves markedly towards neutral when the virus is propagated in cat kidney, as opposed to mouse mammary gland cells, demonstrating a strong host-cell influence presumably on the composition of the carbohydrate moieties (NUSSE et al. 1980). Isotopic labelling experiments and direct chemical analysis show that both MMTV glycoproteins carry complex carbohydrates containing mannose, fucose, galactose, glucosamine, and sialic acid (YAGI et al. 1978; DICK-

SON and ATTERWILL 1980). A likely explanation for the charge heterogeneity of the two proteins is a variability in content of sialic acid, commonly the terminal sugar residue on carbohydrate side chains (SCHLOEMER et al. 1976), since prior treatment of the glycoproteins with neuraminidase reduces the observed heterogeneity (NUSSE et al. 1980).


## 4.2  Intracellular Precursors for *env*

Its occurrence on the external surface of the virion, its abundance, and its readiness to radiolabel have made gp52 very amenable to detailed immunological studies. Consequently, gp52 has formed the basis of several radioimmunoassays (CARDIFF 1973; PARKS et al. 1974; VERSTRAETEN et al. 1975; RITZI et al. 1976; SHEFFIELD et al. 1977; ZARGERLE et al. 1977) and proved to be one of the principal indicators of polymorphism among different strains of MMTV (TERAMOTO et al. 1977b; ARTHUR et al. 1978a; TERAMOTO and SCHLOM 1979; MARCUS et al. 1979; CALBERG-BACQ et al. 1981). One of the major uses of monospecific antisera to gp52 has been the dissection of the intracellular events involved in the synthesis and maturation of the MMTV glycoproteins. Immunoprecipitation of labelled cell extracts identifies a 73000-dalton polyprotein as the major *env*-related product (Fig. 3), and subsequent tryptic peptide mapping has confirmed that this putative precursor, $Pr73^{env}$, contains both gp52 and gp36 (DICKSON et al. 1976; NUSSE et al. 1978; RACEVSKIS and SARKAR 1978; SCHOCHETMAN et al. 1978; ANDERSON et al. 1979). From experiments using hypotonic shock to synchronize the initiation of protein synthesis (SCHOCHETMAN et al. 1977) or pactamycin inhibition of initiation (DICKSON and ATTERWILL 1980; SEN et al. 1980b) as described previously for the *gag* gene products, the order of the proteins in the precursor was established as $NH_2$–gp52–gp36–COOH (Fig. 3), a conclusion that has since been verified by direct sequencing of the *env* gene (Fig. 4; REDMOND and DICKSON 1983).

Labelling experiments with radioactive sugars indicate that $Pr73^{env}$ is glycosylated, the predominant sugars being mannose and glucosamine (ANDERSON et al. 1979; ANDERSON and NASO 1980; DICKSON and ATTERWILL 1980). This is consistent with many other studies on glycoproteins which have shown that cores of sugar residues, consisting mainly of mannose and glucosamine, are assembled on a lipid carrier molecule and added as units to the nascent polypeptide chain during synthesis. Carbohydrate addition occurs almost exclusively at asparagine residues, where the amino acid sequence is Asp-X-Thr/Ser (PLESS and LENNARZ 1977), but can be specifically blocked by the antibiotic tunicamycin (TAKATSUKI et al. 1975; LEAVITT et al. 1977). Thus, MMTV-producing cells labelled in the presence of tunicamycin accumulate a non-glycosylated form of the *env* precursor with a molecular mass of 60000 daltons (Fig. 3; DICKSON and ATTERWILL 1980; ARTHUR et al. 1982). Conversely, a similar apoprotein can be obtained by

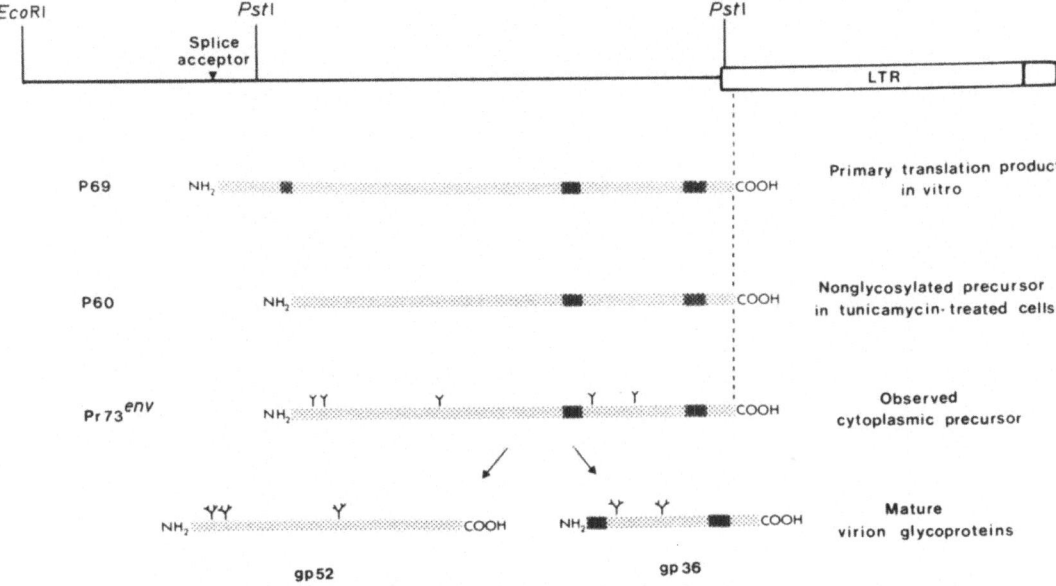

**Fig. 3.** Expression of the MMTV envelope gene. The region of the DNA provirus which encodes the envelope products is shown with the cleavage sites for the restriction enzymes *Eco*RI and *Pst*I indicated. The protein products of the envelope gene under various in vitro and in vivo conditions are depicted by the *stipled bars*. Darker *shading* identifies regions of uncharged amino acids. The sites of primary glycosylation (Y) are predicted from the DNA sequence (Fig. 4). These undergo secondary glycosylation during the maturation of the virion glycoproteins

removing the bulk of the core sugars from $Pr73^{env}$ with the enzyme *N*-acetyl-$\beta$-glucosaminidase H (DICKSON and ATTERWILL 1980; ARTHUR et al. 1982). This enzyme cleaves off the primary cores of oligosaccharides as units, leaving a single glucosamine residue linked to the asparagine in the polypeptide chain (TARENTINO and MALEY 1974). By using suboptimal conditions for enzymic digestion or lower concentrations of tunicamycin, partially glycosylated forms of the *env* precursor can be generated, differing in molecular weight according to the number of core sugar units present. Such experiments identify five discrete forms of the *env* precursor, evenly distributed in the molecular mass range of 60000 to 73000 daltons, suggesting the addition of five carbohydrate side chains to the precursor during its synthesis (DICKSON and ATTERWILL 1980). This is consistent with the derived amino acid sequence of *env*, which predicts five potential attachment sites (see Figs. 3, 4; REDMOND and DICKSON 1983) corresponding to the Asp-X-Thr/ Ser format.

In vivo labelling of infected cells also reveals a slightly larger form of $Pr73^{env}$, which contains additional sugar moieties, and possibly represents an abnormal form of the precursor which accumulates on the cell surface

```
1                                                                           18
MET Pro Asn His Gln Ser Gly Ser Pro Thr Gly Ser Ser Asp Leu Leu Leu Ser
                                                                            36
Gly Lys Lys Gln Arg Pro His Leu Ala Leu Arg Arg Lys Arg Arg Arg Glu MET
                                                                            54
Arg Lys Ile Asn Arg Lys Val Arg Arg MET Asn Leu Ala Pro Ile Lys Glu Lys
                                                                            72
Thr Ala Trp Gln His Leu Gln Ala Leu Ile Ser Glu Ala Glu Glu Val Leu Lys
                                                                            90
Thr Ser Gln Thr Pro Gln Asn Ser Leu Thr Leu Phe Leu Ala Leu Leu Ser Val
                                                                           108
Leu Gly Pro Pro Pro Val Thr Gly Glu Ser Tyr Trp Ala Tyr Leu Pro Lys Pro
                                                                           126
Pro Ile Leu His Pro Val Gly Trp Gly Ser Thr Asp Pro Ile Arg Val Leu Thr
 ◆                                                                ◆        144
Asn Gln Thr MET Tyr Leu Gly Gly Ser Pro Asp Phe His Gly Phe Arg Asn MET
                                                                           162
Ser Gly Asn Val His Phe Glu Gly Lys Ser Asp Thr Leu Pro Ile Cys Phe Ser
                                                                           180
Phe Ser Phe Ser Thr Pro Thr Gly Cys Phe Gln Val Asp Lys Gln Val Phe Leu
                                                                           198
Ser Asp Thr Pro Thr Val Asp Asn Asn Lys Pro Gly Gly Lys Gly Asp Lys Arg
                                                                           216
Arg MET Trp Glu Leu Trp Leu His Thr Leu Gly Asn Ser Gly Ala Asn Thr Lys
                                                                           234
Leu Val Pro Ile Lys Lys Lys Leu Pro Pro Lys Tyr Pro His Cys Gln Ile Ala
                                                                           252
Phe Lys Lys Asp Ala Phe Trp Glu Gly Asp Glu Ser Ala Pro Pro Arg Trp Leu
                                                                           270
Pro Cys Ala Phe Pro Asp Lys Gly Val Ser Phe Ser Pro Lys Gly Ala Leu Gly
                                                                           288
Leu Leu Trp Asp Phe Ser Leu Pro Ser Pro Ser Val Asp Gln Ser Asp Gln Ile
                           ◆                                               306
Lys Ser Lys Lys Asp Leu Phe Gly Asn Tyr Thr Pro Pro Val Asn Lys Glu Val
                                                                           324
His Arg Trp Tyr Glu Ala Gly Trp Val Glu Pro Thr Trp Phe Trp Glu Asn Ser
                                                                           342
Pro Lys Asp Pro Asn Asp Arg Asp Phe Thr Ala Leu Val Pro His Thr Glu Leu
                                                                           360
Phe Arg Leu Val Ala Ala Ser Arg His Leu Ile Leu Lys Arg Pro Gly Phe Gln
                                                                           378
Glu His Glu MET Ile Pro Thr Ser Ala Cys Val Thr Tyr Pro Tyr Ala Ile Leu
                                                                           396
Leu Gly Leu Pro Gln Leu Ile Asp Ile Glu Lys Arg Gly Ser Thr Phe His Ile
                                                                           414
Ser Cys Ser Ser Cys Arg Leu Thr Asnys Leu Asp Ser Ser Ala Tyr Asp Tyr
                                                                           432
Ala Ala Ile Ile Val Lys Arg Pro Pro Tyr Val Leu Leu Pro Val Asp Ile Gly
                                                                           450
Asp Glu Pro Trp Phe Asp Asp Ser Ala Ile Gln Thr Phe Arg Tyr Ala Thr Asp
                                                                           468
Leu Ile Arg Ala Lys Arg Phe Val Ala Ala Ile Ile Leu Gly Ile Ser Ala Leu
                                                                           486
Ile Ala Ile Ile Thr Ser Phe Ala Val Ala Thr Thr Ala Leu Val Lys Glu MET
                               ◆                                           504
Gln Thr Ala Thr Phe Val Asn Asn Leu His Arg Asn Val Thr Leu Ala Leu Ser
                                                                           522
Glu Gln Arg Ile Ile Asp Leu Lys Leu Glu Ala Arg Leu Asn Ala Leu Glu Glu
                                                                           540
Val Val Leu Glu Leu Gly Gln Asp Val Ala Asn Leu Lys Thr Arg MET Ser Thr
                                                                     ◆     558
Arg Cys His Ala Asn Tyr Asp Phe Ile Cys Val Thr Pro Leu Pro Tyr Asn Ala
                                                                           576
Thr Glu Asp Trp Glu Arg Thr Arg Ala His Leu Leu Gly Ile Trp Asn Asp Asn
                                                                           594
Glu Ile Ser Tyr Asn Ile Gln Glu Leu Thr Asn Leu Ile Ser Asp MET Ser Lys
                                                                           612
Gln His Ile Asp Ala Val Asp Leu Ser Gly Leu Ala Gln Ser Phe Ala Asn Gly
                                                                           630
Val Lys Ala Leu Asn Pro Leu Asp Trp Thr Gln Tyr Phe Ile Phe Ile Gly Val
                                                                           648
Gly Ala Leu Leu Leu Val Ile Val Leu MET Ile Phe Pro Ile Val Phe Gln Cys
                                                                           666
Leu Ala Lys Ser Leu Asp Gln Val Gln Ser Asp Leu Asn Val Leu Leu Leu Lys
                                                                           684
Lys Lys Lys Gly Gly Asn Ala Ala Pro Ala Ala Glu MET Val Glu Leu Pro Arg

Val Ser Tyr Thr .
```

**Fig. 4.** Amino acid sequence for the envelope region of the GR strain of MMTV. The methionine codons which might serve as potential initiation sites are *underlined* and the known amino terminal ends of gp52 and gp36 are indicated by *square brackets*. *Shaded* areas underscore regions of uncharged amino acids, and sites of primary glycosylation are denoted as ◆

(ANDERSON et al. 1979; DICKSON and ATTERWILL 1980). Another larger
MMTV-related glycoprotein of 130000 daltons has been reported but again
its function and nature remain obscure (ANDERSON and NASO 1980).

## 4.3  Expression of the *env* Gene In Vivo and In Vitro

In terms of structure, location, and glycosylation, the MMTV *env* proteins
appear to conform to the general pattern established for other viral and
cellular glycoproteins. This analogy is further strengthened by the demonstration that Pr73$^{env}$ is synthesized in the rough endoplasmic reticulum,
probably as a transmembrane protein (DICKSON and ATTERWILL 1980). The
weight of evidence to date indicates that transmembrane synthesis of glycoproteins is mediated by a short, hydrophobic "signal sequence" at the amino
terminus of the primary translation product (see AUSTEN 1979). As this
leader peptide appears to be removed by a specific protease before the
nascent polypeptide is complete, it is not detected in cell extracts under
normal labelling conditions. Its presence can, however, be demonstrated
in cell-free translation systems which lack membranes, and, therefore, do
not process the signal peptide or have the capacity to glycosylate proteins.
Thus, in vitro translation of MMTV RNA generates an *env*-related polyprotein of 67000 − 69000 daltons (SEN et al. 1979, 1981; DICKSON and PETERS
1981; ARTHUR et al. 1982; DUDLEY and VARMUS 1981). Relative to the
60000-dalton non-glycosylated precursor detected in tunicamycin-treated
cells, this primary translation product would appear to have an amino terminal leader peptide of about 7000–9000 daltons (Fig. 3). Although a similar
result has been reported for the avian leukosis virus *env* product, this
7000–9000-dalton leader is unusually long, the more typical size being 20–30
amino acids, so that the possibility remains open for an as yet unspecified
function for this extra long sequence.

Synthesis of this 69000-dalton precursor can be achieved using both
viral genome RNA and intracellular mRNA. In the first instance, MMTV
genomic RNA was isolated from mature virions and partially degraded
in order to expose internal methionine codons which might have the potential for initiating protein synthesis. By size-selecting polyadenylated fragments of this degraded genome RNA, a class of molecules can be prepared
which programme the synthesis of a series of non-glycosylated *env*-related
proteins, the largest of which is 69000-daltons (DICKSON and PETERS 1981).
These experiments have the additional bonus of verifying the gene order
in the MMTV genome as 5′–*gag–pol–env*–3′. In intact viral RNA, therefore,
the initiation codon for the primary product of the *env* gene is internal
and presumably inaccessible. In vivo, the normal route for expression of
the *env* gene is via a subgenomic mRNA in which sequences from the 5′
end of the viral genome are spliced to the body of the *env* gene, circumvent-
ing *gag* and *pol*. Thus, MMTV-infected cells contain a 24–S subgenomic
RNA in addition to the full-length 35 S genome RNA. Purification and

in vitro translation of this 24 S mRNA also yields a 69000-dalton *env*-related protein (SEN et al. 1979; DUDLEY and VARMUS 1981; ROBERTSON and VARMUS 1981; ARTHUR et al. 1982).

## 4.4 Amino Acid Sequence of the *env* Gene Products

In recent years, the application of recombinant DNA technology and the development of rapid DNA sequencing methods have contributed enormously to our understanding of retrovirus systems and in some cases have permitted the derivation of the complete nucleotide sequence of the viral genome. With MMTV, more modest but considerable progress has been made, such that we now know the sequence of some 4000 nucleotides extending from the single *Eco*RI restriction site to the 3' end of the proviral DNA (Fig. 3). This sequence includes a major continuous open reading frame for 688 amino acids which encompasses the entire *env* gene (REDMOND and DICKSON 1983). By correlating the predicted amino acid sequence of this region with the directly determined terminal sequences of gp52 and gp36, as shown in Fig. 4, it has been possible to draw a number of further conclusions regarding the organization of the *env* gene and the properties of the two MMTV glycoproteins (REDMOND and DICKSON 1983).

1. The known amino terminus of gp52 (ARTHUR et al. 1982), when aligned with the predicted sequence shown in Fig. 4 (position 99), agrees at 41 out of 43 residues. The two differences can be accounted for by two single base changes and probably reflect strain differences between the viruses used in the determinations.
2. The amino terminus of gp52 is immediately preceded by a region (positions 80 to 98) of 19 uncharged amino acids, which may constitute a signal sequence for transmembrane synthesis. Cleavage of the signal peptide would therefore take place at a small amino acid residue, in this case glycine, in agreement with observations from other glycoproteins (see AUSTEN 1979). Normally such a signal sequence would occur at the amino terminus of the primary translation product. However, this is clearly not the case with MMTV (or avian sarcoma virus; SCHWARTZ et al. 1982), where the signal sequence appears to be only part of a much longer leader.
3. The start of gp52 is preceded by three in-frame methionine codons (positions 1, 36, and 46), any one of which could serve as the initiation site for the primary *env* product. The sizes of the predicted leader peptides, 11000, 7000, and 5700 daltons, respectively, are consistent with previous estimates for the leader, based on SDS-acrylamide gel electrophoresis, but are sufficiently close to preclude any decisions as to which methionine codon represents the true start of *env*. Moreover, it is possible that the true initiation site might be provided by sequences from the 5' region of the genome, linked to *env* by RNA splicing, as is the case in avian

sarcoma virus (E. Hunter and D. Schwartz, personal communication). The potential splice point for generating the 24 S *env* mRNA has been mapped immediately upstream (within a few nucleotides) of the first of these methionine codons.

4. The known amino terminus of gp36 (HENDERSON et al. 1983) can be aligned with the sequence (starting at position 457) and agrees exactly over 27 residues. The four amino acids immediately preceding this position, –Arg–Ala–Lys–Arg–, constitute a typical cleavage site for a protease with a specificity similar to trypsin. Analogous cleavage sites have been found between other viral glycoproteins (PORTER et al. 1979; SCHWARTZ et al. 1982; SHINNICK et al. 1981; LENZ et al. 1982).

5. The predicted amino acid sequence of gp36 reveals two extensive regions of hydrophobicity comprising 27 and 30 amino acids, respectively. The first of these is located at the amino terminus of gp36 and, although interaction with the envelope membrane cannot be excluded, parallels drawn from other viral systems would suggest that this hydrophobic element is involved in the association between gp36 and gp52 and that, when exposed, it may function in membrane fusion (WHITE et al. 1981). Also by analogy with other surface glycoproteins, it seems probable that the second hydrophobic region (positions 621 and 650) represents the transmembrane segment of gp36. This would leave 38 amino acids on the inner side of the membrane which could be involved in interactions with some of the viral core proteins.

6. The carboxy terminus of gp36, as determined by direct protein sequence analysis (HENDERSON et al. 1983), has been located in the predicted sequence at position 689 and agrees exactly over the final five residues. The most interesting aspect of this finding is that the end of the *env* gene must, therefore, overlap the proviral LTR by 53 nucleotides and must also encompass the polypurine tract preceding the LTR which is thought to be involved in priming the synthesis of positive-strand viral DNA (MAJORS and VARMUS 1981). This is reflected in the unusual sequence –Lys–Lys–Lys–Lys (position 666 to 670) in the cytoplasmic tail of gp36.

7. Five potential sites for carbohydrate addition are predicted by the sequence, three in gp52 and two in gp36 (position 127, 143, 297, 498, and 557), in exact agreement with previous experimental observations.

8. The early experiments involving iodination of intact virions showed preferential labelling of gp52, with little or no labelling of gp36 (CARDIFF 1973; WITTE et al. 1973). A partial explanation for this is provided by the observation that gp52 contains 13 tyrosine residues, of which two may be rendered inaccessible by the proximity of carbohydrate side chains, compared with only three tyrosine residues in gp36, of which one is in the transmembrane segment.

9. One final conclusion from the nucleotide sequence is that the organization of the MMTV genome may be very conservative in that two other open reading frames are observed which overlap with the *env* gene, albeit in different frames. The first of these extends for approximately 800 nucleo-

tides from the *Eco*RI site in the provirus and is presumed to represent the carboxy terminus of the *pol* gene. The second begins at the 5′ boundary of the right-hand LTR of the  provirus and, as will be discussed in the following section, extends for about 960 nucleotides within the U3 segment of the LTR and codes for a series of non-structural proteins for which a function has yet to be determined.

# 5   Mouse Mammary Tumour Virus *orf* Proteins

## 5.1   Discovery of *orf*

One unusual feature of  MMTV that has only recently come to light is that it has the potential to encode what may represent an additional gene or gene product, quite distinct from the structural genes *gag*, *pol*, and *env*. However, since no function has yet been ascribed to these sequences or the proteins which they encode, it is still not clear whether they can be defined as a gene; they are, therefore, referred to simply as an open reading frame, abbreviated as *orf*. Evidence for the existence of this open reading frame came from two independent approaches: in vitro protein synthesis and DNA sequencing.

As indicated previously, MMTV viral RNA is in the positive sense and can serve as a messenger RNA both in vivo and in vitro. Although some in vitro translation systems appear to be relatively promiscuous, initiation of protein synthesis is generally restricted to the methionine codon closest to the 5′ end of the template RNA. Thus, the principal proteins synthesized from full-length MMTV RNA are the *gag*-related products (DAHL and DICKSON 1979; SEN et al. 1979). However, by partially degrading the viral RNA, internal methionine codons can be exposed which may act as potential initiation sites, When such a strategy was used in combination with polyA selection and size fractionation of the resultant RNA fragments, a surprising result was obtained. As well as the clearly recognizable products of the *gag*, *pol*, and *env* genes, a series of four methionine-rich proteins were detected whose peptide maps are completely unrelated to any of the known viral proteins (DICKSON and PETERS 1981; SEN et al. 1981). Their peptide maps also represent an overlapping series, suggesting that the four products probably reflect initiation at different methionine codons within the same open reading frame (Fig. 5). Although the smaller 18000-, 21000-, and 24000-dalton species tend to be the predominant in vitro products, the largest of the so-called *orf* proteins, with a molecular mass of 36000 daltons, would require a continuous open reading frame of almost 1000 nucleotides.

From the size of the RNA fragments which encode the *orf* proteins, the sequences must map close to the 3′ end of the viral genome. This consideration, plus the fact that the MMTV provirus has unusually long LTR segments, raised the possibility that *orf* may map within the LTR, a conclu-

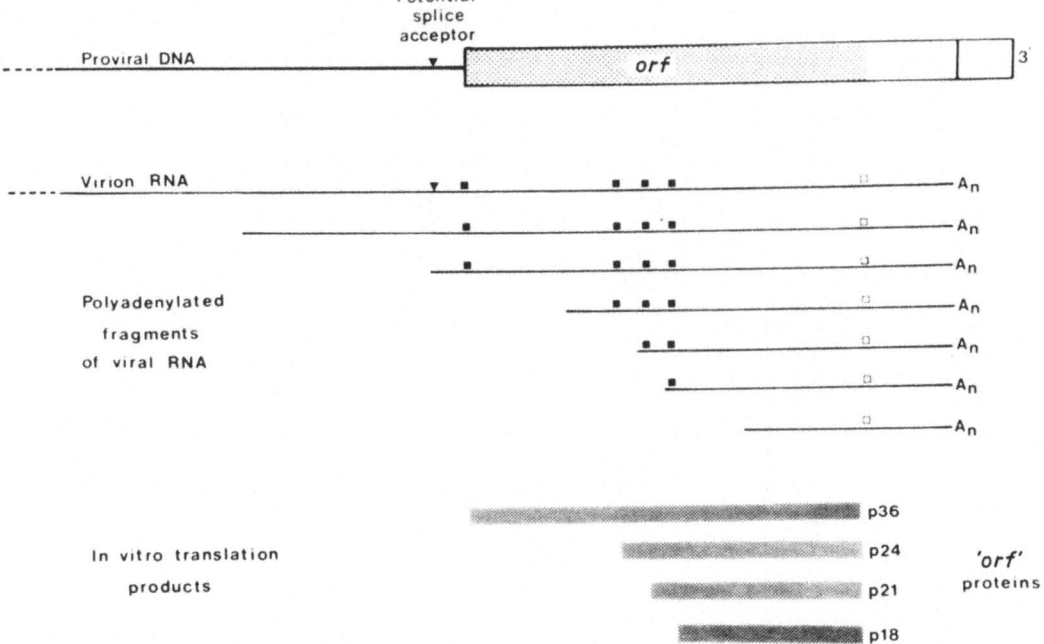

**Fig. 5.** Organization of the MMTV *orf* sequences. Schematic representation of the principle used to detect the *orf* proteins and the region of the viral genome from which they derive. Fragments of virion RNA less than about 1500 nucleotides in length were found to be capable of directing the synthesis of a series of related proteins quite distinct from any known virion proteins. Similar proteins were detected by in vitro expression of cloned LTR DNA. The methionine codons used for the initiation of protein synthesis in vitro are indicated by ■ and the termination codon in the same frame by □. The *orf* proteins and their size designation are shown

sion which was subsequently verified by the coupled in vitro transcription and translation of LTR DNA cloned into bacterial plasmids (DICKSON et al. 1981). Similar conclusions were arrived at independently by determining the DNA sequence of the cloned MMTV LTRs (DONEHOWER et al. 1981; FASEL et al. 1982; KENNEDY et al. 1982). These studies indicated that a continuous open reading frame did indeed exist, beginning at the 5′ boundary of the LTR and extending for about 960 nucleotides. Examination of the DNA sequence identified internal methionine codons at which the smaller *orf* proteins are presumably initiated in the in vitro system but revealed that the adenosine of the (ATG) initiation codon for the 36000-dalton species actually lies outside the LTR. Since this A residue forms part of the polypurine tract implicated in positive-strand DNA synthesis (and, as discussed earlier, is also part of the *env* gene), it is invariably present in the viral RNA and in the sequence adjacent to the 3′ LTR of the provirus. However, depending on the sequence of cellular DNA at the site of integration, it would not necessarily occur adjacent to the 5′ LTR. The implication

```
                                                      Leu
MET Pro Arg Leu Gln Gln Lys Trp Leu Asn Ser Arg Glu Cys Pro Thr Pro Arg
                                                      Leu

Gly Glu Ala Ala Lys Gly Leu Phe Pro Thr Lys Asp Asp Pro Ser Ala His Lys
Arg                                                          Cys Thr
    MET                                 Leu
Arg Val Ser Pro Ser Asp Lys Asp Ile Phe Ile Leu Cys Cys Lys Leu Gly Ile
    MET                                 Leu

Ala Leu Leu Cys Leu Gly Leu Leu Gly Glu Val Ala Val Arg Ala Arg Arg Ala

                    ⌈Asn Asn⌉                                        Asp
Leu Thr Leu Asp Ser Phe Asn Ser Ser Ser Val Gln Asp Tyr Asn Leu Asn Asn
                        Asn
                            Gly
Ser Glu Asn Ser Thr Phe Leu Leu Arg Gln Gly Pro Gln Pro Thr Ser Ser Tyr
                            Gly
        Ile           Leu                                           Lys
Lys Pro His Arg Phe Cys Pro Ser Glu Ile Glu Ile Arg MET Leu Ala Lys Asn
        Leu

Tyr Ile Phe Thr Asn Lys Thr Asn Pro Ile Gly Arg Leu Leu Val Thr MET Leu
                                                        Ile MET

Arg Asn Glu Ser Leu Ser Phe Ser Thr Ile Phe Thr Gln Ile Gln Lys Leu Glu
                                                          Arg
                                    Lys
MET Gly Ile Glu Asn Arg Lys Arg Arg Ser Thr Ser Ile Glu Glu Gln Val Gln
                                        Val
        Ala Ser
Gly Leu Leu Thr Thr Gly Leu Glu Val Lys Lys Gly Lys Lys Ser Val Phe Val
        Arg Ala Ser              Arg          Arg     Thr Leu
                                Pro Arg
Lys Ile Gly Asp Arg Trp Trp Gln Leu Gly Thr Tyr Arg Gly Pro Tyr Ile Tyr
                                Pro

Arg Pro Thr Asp Ala Pro Leu Pro Tyr Thr Gly Arg Tyr Asp Leu Asn Trp Asp
                                                              Phe
                Ile
Arg Trp Val Thr Val Asn Gly Tyr Lys Val Leu Tyr Arg Ser Leu Pro Phe Arg

                                        Thr Glu Lys
Glu Arg Leu Ala Arg Ala Arg Pro Pro Trp Cys MET Leu Ser Gln Glu Glu Lys
                                            Val     Thr

Asp Asp MET Lys Gln Gln Val His Asp Tyr Ile Tyr Leu Gly Thr Gly MET His
    Ile                                                          Asn
            Val                             Ala
Phe Trp Gly Lys Ile Phe His Thr Lys Glu Gly Thr Val Ala Gly Leu Ile Glu
Val                    ⌊Tyr⌋               Ala         Arg Gln Leu
        Ala                         Asp
His Tyr Ser Pro Lys Thr Tyr Gly MET Ser Tyr Tyr Glu
    Ile     Ala Asp     Phe               Asn Gly
```

**Fig. 6.** Amino acid sequence of the postulated *orf* proteins as predicted from the DNA sequences of cloned MMTV LTRs. The continuous sequence was derived from the C3H strain of MMTV (DONEHOWER et al. 1981; J. Majors personal communication) with differences detected in the GR strain (*above*) (FASEL et al. 1982) and endogenous unit II (KENNEDY et al. 1982) (*below*)

is that if the 36000-dalton *orf* protein is indeed synthesized in vivo, it is more likely to be encoded by a mRNA derived from the 3′ end of the genome, although clearly other mechanisms cannot be excluded. The general absence of such a potential mRNA in MMTV-producing cell lines and tumour tissue has until recently cast doubt on the functional significance of the *orf* sequences. However, two laboratories now report the detection of an approximately 1.6–kb mRNA species (WHEELER et al. 1983; VAN OOYEN, personal communication), specifically in outgrowth lines derived from hyperplastic alveolar nodules, the pre-neoplastic lesions commonly found in the lactating mammary glands of certain strains of mice. This

RNA appears to consist of sequences from the 5' end of the viral genome spliced to the *orf* sequences at a site approximately 80 bases upstream from the boundary of the LTR (Fig. 5). A consensus splice acceptor site has been located at the appropriate position in the DNA sequence of the *env* gene (REDMOND and DICKSON 1983).

These recent observations encourage the belief that the *orf* proteins may have an in vivo role. Such a view gains support from the fact that the *orf* sequences are conserved by different strains of MMTV, both exogenously and endogenously transmitted. For example, Fig. 6 shows the amino acid sequence of the 36000-dalton *orf* protein from the C3H strain of virus (DONEHOWER et al. 1981; J. Majors, personal communication), with the differences in the corresponding proteins from the GR strain (FASEL et al. 1982) and the endogenous proviral unit II (KENNEDY et al. 1982) depicted above and below. Similar conclusions were derived from in vitro expression studies using a variety of cloned LTRs (PETERS et al. 1982), the significant point being that, although a number of differences could be detected in the DNA and amino acid sequences, all MMTV strains examined so far have maintained a continuous open reading frame.

## 5.2  Possible Functions for *orf*

Unfortunately, little progress in assessing the function of *orf* can be made without a suitable assay for the presence or activity of the proteins in vivo. Attempts are underway in a number of laboratories to prepare antisera directed either against selected synthetic peptides (predicted from the DNA sequence) or against larger segments of the *orf* proteins synthesized in bacteria. Until such reagents provide a greater insight, we are left to speculate as to the possible roles for *orf*. For example, the detection of a mRNA in hyperplastic tissue might suggest that the MMTV *orf* product(s) acts as a mitogen or growth factor, inducing cells to proliferate more rapidly. In this context, *orf*, though not a true oncogene, may contribute to the oncogenic event by rendering cells more vulnerable to some other stimulus. Alternatively, a role in integration or transposition might be envisaged in view of the strong structural similarities between the provirus and the transposable elements of bacteria (VARMUS 1982 for review). However, it should be noted that the same similarity to transposons exists for all retroviruses, yet MMTV is the only one to date in which the LTR has been found to have any significant protein-coding capacity. The other obvious feature unique to MMTV is the sensitivity of viral gene expression to modulation by glucocorticoids. Although some correlation between *orf* and hormone responsiveness would seem to be an attractive proposition, recent experiments aimed at defining the elements responsible for hormonal regulation have shown that the entire open reading frame can be deleted from the LTR without influencing steroid sensitivity. (RINGOLD, this volume).

These speculations represent only a few of the many possible functions that might be envisaged for *orf*, their only merit being in suggesting directions for future experimentation. It will be interesting to see what such experiments eventually reveal about the true role, if any, of the MMTV *orf* proteins.


# 6  Tissue and Species Diversity

## 6.1  Mouse Mammary Tumour Virus Expression in Other Tissues

So far in this review, we have, wherever possible, discussed MMTV as if it were a single, discrete virus, replicating in mouse mammary gland cells, and have intentionally stressed unifying principles rather than cloud the issue with sometimes contradictory details. In this section we intend to go some way toward redressing the balance by drawing attention both to the existence of multiple strains of MMTV and to evidence for its expression in diverse cell types and species.

The most common natural route of infection for MMTV is via the milk of viremic females. As a consequence, recipient mice are exposed to virus during suckling at a time when their mammary gland is hardly recognizable as a distinct tissue or organ. These considerations have led to the search for alternative tissues in which infection may become established and which may therefore act as a source of virus for subsequent infection of the mammary gland, perhaps at puberty. Such studies principally rely on biological activity, such as tumour induction, or on the detection of viral antigens in fresh or fixed tissue samples, for example, by immunofluorescence or radioimmunoassay (RIA) procedures. Although no obvious route of infection has been established and despite some variability in the results, it is now clear that tissues other than the mammary gland show biological activity and may or may not express MMTV antigens. The bioactive tissues include lymphoid cells from several organs, the salivary gland, kidney, and male accessory sex glands (see NANDI and MCGRATH 1973; HILGERS and BENTVELZEN 1978 for extensive reviewing).

Perhaps the most striking result has been the high level of MMTV antigen expression in lymphocytes. Curiously, in most cases of lymphoid cell lines where MMTV expression has been reported, there appears to be some defect in processing of the intracellular precursors for the viral proteins, resulting, for example, in the detection of Pr73$^{env}$, or some aberrant form of it, on the cell surface (NUSSE et al. 1979; VAIDYA et al. 1980; RACEVSKIS and SARKAR 1982). Lymphoid cells also contain a large number of intracytoplasmic A-type particles, to the extent that they have been exploited as a preparative source for such particles (TANAKA et al. 1972; SMITH and WIVEL 1973; SARKAR and WHITTINGTON 1977). Again, these particles appear to contain unprocessed or possibly aberrantly processed forms of the *gag* precursor. However, this latter phenomenon is not unique to lymphoid cells, and it seems likely that all naturally occurring intracytoplasmic A-type particles represent viral cores assembled with the unprocessed or partially processed precursor for the internal structural proteins (TANAKA 1977; SMITH 1978; RACEVSKIS et al. 1981). Thus, A-type particles isolated in the presence of protease inhibitors contain a single, major, approximately 70000-dalton, *gag*-related protein.

A similar block in the maturation of the MMTV proteins may in part explain the apparent lack of infectivity of the virus in tissue culture systems. It is even conceivable that the mammary epithelium represents one of the few infectable tissues in which an optimal level of processing can be achieved. However, the high levels of MMTV antigens in lymphoid cells may have implications beyond the aberrant processing of the viral proteins, in that rearrangements and/or amplification of MMTV proviruses have also been reported in some mouse lymphomas (MICHALIDES et al. 1982).

## 6.2 Diversity of Mouse Mammary Tumour Virus in Different Strains and Species

The search for MMTV components in various tissues and body fluids has relied almost exclusively on the availability of specific antisera and the ability to make quantitative assessments by competition RIAs. Not surprisingly, the first MMTV protein to be exploited in this way was gp52, since it can be readily purified and is the only protein which can be labelled by radio-iodination on intact virions (CARDIFF 1973; PARKS et al. 1974; VERSTRAETEN et al. 1975; RITZI et al. 1976; SHEFFIELD et al. 1977; ZARGERLE et al. 1977). However, RIAs have now been developed for a number of the internal proteins, including p27, p14, and p10 (PARKS et al. 1974; NOON et al. 1975; ARTHUR et al. 1978a; HENDRICK et al. 1978; ARTHUR and FINE 1979). Clearly, the specificity of the various RIAs can be refined by the selective adsorption of antisera to enhance, for example, detection of type-specific, as opposed to group-specific, determinants. Thus, as well as confirming the high degree of cross-reactivity among MMTVs from various inbred mouse strains, RIAs were instrumental in delineating antigenic differences among the various strains. In assays directed against either of the MMTV glycoproteins, both group-specific and type-specific determinants can be detected for the MMTVs of the C3H, R111, GR, C3Hf, and MB$^+$ Swiss strains of mice (TERAMOTO et al. 1977b; ARTHUR et al. 1978a; TERAMOTO and SCHLOM 1979; MARCUS et al. 1979; CALBERG-BACQ et al. 1981). Type-specific determinants have also been reported to occur on the major internal protein, p27 (TERAMOTO and SCHLOM 1978; CALBERG-BACQ et al. 1981). More recently, monoclonal antibody techniques have been employed, and the availability of such reagents should prove very useful in further dissection of the various antigenic and functional domains on the MMTV proteins (MASSEY et al. 1980; COLCHER et al. 1981).

The ubiquity and high degree of relatedness of MMTVs in the germ lines of most, if not all, laboratory strains of mice led some investigators to look for the same or related viruses in other species. One positive example is the Asian rodent, *Mus cervicolor,* which was found to harbour a milk-borne virus particle morphologically and biochemically similar to the MMTV of *Mus musculus* (SCHLOM et al. 1978; HORAN-HAND et al. 1980). Competition RIAs for both gp52 and p27 demonstrated some, but only partial, cross-reactivity between the proteins of the new virus, known as MC-MTV, and those of MMTV. By establishing an interspecies RIA utilizing antiserum against MC-MTV and labelled glycoproteins (either gp52 or gp36) from MMTV, it is possible to detect related viral antigens in two additional species, *Mus cookii* (MCo-MTV) and *Mus caroli* (MCa-MTV) (TERAMOTO et al. 1980). Thus, it is becoming apparent that the MMTVs may form only part of a much wider family of viruses present in the genus *Mus.* Some of these may even be present in *Mus musculus,* as nucleic acid hybridization at low stringency has revealed the presence in the germ line of sequences related to, but quite distinct from, those of the now well-characterized MMTV provirus (CALLAHAN et al. 1982a). Curiously, a similar, but only distantly related, set of sequences has also been detected in the human genome (CALLAHAN et al. 1982b). This phenomenon may conceivably account for the numerous, but somewhat variable, reports of cross-reactivity between the MMTV gp52 and p27 and sera from human patients (WEISS et al. 1982). The significance of these findings obviously remains obscure, but the possibility that MMTV represents only one well-defined branch of a large and widely distributed family of viruses should not be discounted.


# 7 Concluding Remarks


From the length and scope of this review, it is clear that we now have a fairly comprehensive picture of the proteins encoded by MMTV. Nevertheless, there are obvious and substantial gaps in our knowledge which will require considerable future commitment to fill. Outstanding among these are the mysteries surrounding the synthesis of $Pr110^{gag}$ and $Pr160^{gag-pol}$, the significance of p30 and p8, and the details surrounding the processing and function of the individual *gag* proteins. A major effort in cloning and sequencing of the DNA may answer many, but clearly not all, of these questions. Even though we now have a complete sequence for the *env* gene, we still do not know which methionine codon is used for initiation of protein synthesis, what the role of the long leader peptide might be, and, more importantly, how the envelope glycoproteins function in penetrating the host cell and virus assembly. A similar, but even less clear, situation exists in the case of *orf,* where, although the DNA sequence is known, no real clues as to the functional significance of *orf* have been forthcoming. However, given the current rate of advancement in our understanding of MMTV, we can only be optimistic that the future holds great promise for unravelling the secrets of this intriguing virus.

# References

Anderson SJ, Naso RB (1980) A unique glycoprotein containing GR-mouse mammary tumor virus peptides and additional peptides unrelated to viral structural proteins. Cell 21:837–847

Anderson SJ, Naso RB, Davis J, Bowen JM (1979) Polyprotein precursors to mouse mammary tumor virus proteins. J Virol 32:507–516

Arthur LO, Fine DL (1979) Immunological characterization of mouse mammary tumor virus p10 and its presence in mammary tumors and sera of tumor-bearing mice. J Virol 30:148–156

Arthur LO, Bauer RF, Orme LS, Fine DL (1978a) Co-existence of the mouse mammary tumor virus (MMTV) major glycoprotein and natural antibodies to MMTV in sera of tumor-bearing mice. Virology 87:266–275

Arthur LO, Long CW, Smith GH, Fine DL (1978b) Immunological characterization of the low molecular weight DNA binding protein of mouse mammary tumor virus. Int J Cancer 22:433–440

Arthur LO, Copeland TD, Oroszlan S, Schochetman G (1982) Processing and amino acid sequence analysis of the mouse mammary tumor virus env gene product. J Virol 41:414–422

Austen BM (1979) Predicted secondary structures of amino-terminal extension sequences of secreted proteins. FEBS Lett 103:308–312

Bernhard W (1958) Electron microscopy of tumor cells and tumor viruses: a review. Cancer Res 18:491–509

Calafat J and Hageman PL (1968) Some remarks on the morphology of virus particles of the B-type and their isolation from mammary tumors. Virology 36:308–312

Calberg-Bacq C-M, Francois C, Kozma S, Osterrieth PM, Teramoto YA (1981) Immunological characterization of a mammary tumor virus from Swiss mice: multiple epitopes associated with the viral gene products. J Gen Virol 57:75–83

Callahan R, Drohan W, Gallahan D, D'Hoostelaere L, Potter M (1982a) Novel class of mouse mammary tumor virus-related DNA sequences found in all species of Mus including mice lacking the virus proviral genome. Proc Natl Acad Sci USA 79:4113–4117

Callahan R, Drohan W, Tronick S, Schlom J (1982b) Detection and cloning of human DNA sequences related to the mouse mammary tumor virus genome. Proc Natl Acad Sci USA 79:5530–5507

Cardiff RD (1973) Quantitation of mouse mammary tumor virus (MMTV) virions by radioimmunoassay. J Immunol 111:1722–1729

Cardiff RD, Young LJT (1980) Mouse mammary tumor biology: An new synthesis. In Viruses in naturally occurring cancers. Essex M, Todaro G, zur Hausen H eds., Cold Spring Harbor Press, New York, pp 1105–1114

Cardiff RD, Puentes MJ, Teramoto YA, Lund JK (1974) Structure of the mouse mammary tumor virus: characterization of bald particles. J Virol 14:1293–1303

Cardiff RD, Puentes MJ, Young LJT, Smith GH, Teramoto YA, Altrock BW, Pratt TS (1978) Serological and biochemical characterization of the mouse mammary tumor virus with localization of p10. Virology 85:157–167.

Colcher D, Horan Hand P, Teramoto YA, Wunderlich D, Schlom J (1981) Use of monoclonal antibodies to define the diversity of mammary tumor viral gene products in virions and mammary tumors of the genus Mus. Cancer Res 41:1451–1459

Dahl H-HM, Dickson C (1979) Cell-free synthesis of mouse mammary tumor virus Pr77 from virion and intracellular mRNA. J Virol 29:1131–1141

Dalton AJ, de Harven E, Dmochowski L, Feldman D, Haguenau F, Harris WW, Howatson AF, Moore D, Pitelka D, Smith K, Uzman B, Zeigel R (1966) Suggestions for the classification of oncogenic RNA viruses. J N C I 37:395–397

Dickson C (1973) Mouse mammary tumour virus RNA-dependent DNA polymerase: requirements and products. J Gen Virol 20:243–247

Dickson C, Atterwill M (1978) Polyproteins related to the major core protein of mouse mammary tumor virus. J Virol 26:660–672

Dickson C, Atterwill M (1979) Composition, arrangement and cleavage of the mouse mammary tumor virus polyprotein precursor Pr77$^{gag}$ and p110$^{gag}$. Cell 17:1003–1012

Dickson C, Atterwill M (1980) Structure and processing of the mouse mammary tumor virus glycoprotein precursor Pr73$^{env}$. J Virol 35:349–361

Dickson C, Peters G (1981) Protein-coding potential of mouse mammary tumor virus genome RNA as examined by in vitro translation. J Virol 37:36–47

Dickson C, Skehel JJ (1974) The polypeptide composition of mouse mammary tumor virus. Virology 58:387–395

Dickson C, Puma JP, Nandi S (1976) Identification of a precursor protein to the major glycoproteins of mouse mammary tumor virus. J Virol 17:275–282

Dickson C, Smith R, Peters G (1981) In vitro synthesis of polypeptides encoded by the long terminal repeat region of mouse mammary tumour virus DNA. Nature 291:511–513

Dion AS, Vaidya AB, Fout GS (1974a) Cation preferences for poly(rC):oligo(dG)-directed DNA synthesis by RNA tumor viruses and human milk particulates. Cancer Res 34:3509–3515

Dion AS, Vaidya AB, Fout GS, Moore DH (1974b) Isolation and characterization of RNA-directed DNA polymerase from a B-type RNA tumor virus. J Virol 14:40–46

Dion AS, Fout GS, Pomenti AA (1979a) In vivo and in vitro phosphorylation of murine mammary tumor virus proteins. J Gen Virol 44:669–678

Dion AS, Pomenti AA, Farwell DC (1979b) Vicinal relationships between the major structural proteins of murine mammary tumor virus. Virology 96:249–257

Donehower LA, Huang AL, Hager GL (1981) Regulatory and coding potential of the mouse mammary tumor virus long terminal redundancy. J Virol 37:226–238

Dudley JP, Varmus HE (1981) Purification and translation of murine mammary tumor virus messenger RNAs. J Virol 39:207–218

Fasel N, Pearson K, Buetti E, Diggelmann H (1982) The region of mouse mammary tumor virus DNA containing the long terminal repeat includes a long coding sequence and signals for hormonally regulated transcription. EMBO J 1:3–7

Feldman SP, Schlom J, Spiegelman S (1973) Further evidence for oncornaviruses in human milk: the production of cores. Proc Natl Acad Sci USA 70:1976–1980

Fleissner E, Tress E (1973) Isolation of a ribonucleoprotein structure from oncornaviruses. J Virol 12:1612–1615

Gautsch JW, Lerner R, Howard D, Teramoto YA, Schlom J (1978) Strain-specific markers for the major structural proteins of highly oncogenic murine mammary tumor viruses by tryptic peptide analyses. J Virol 27:688–699

Gross L (1970) Oncogenic viruses (2nd edn). Pergamon, Oxford

Henderson LE, Sowder R, Smythers G, Oroszlan S (1983) Amino and carboxyl-terminal sequence of mouse mammary tumor virus gp36. J Virol (in press)

Hendrick JC, Francois C, Calberg-Bacq C-M, Colin C, Franchimont P, Gosselin L, Kozma S, Osterrieth PM (1978) Radioimmunoassay for protein p28 of murine mammary tumor virus in organs and serum of mice and search for related antigens in human sera and breast cancer extracts. Cancer Res 38:1826–1831

Hilgers J, Bentvelzen P (1978) Interaction between viral and genetic factors in murine mammary cancer. Adv Cancer Res 26:143–195

Horan-Hand P, Teramoto YA, Callahan R, Schlom WJ (1980) Interspecies radioimmunoassay for the major internal protein of mammary tumor viruses. Virology 101:61–71

Howard DK, Colcher D, Teramoto YA, Young JM, Schlom J (1977) Characterization of mouse mammary tumor virus propagated in heterologous cells. Cancer Res 37:2696–2704

Howk R, Rye L, Killeen L, Scolnick E, Parks W (1973) Characterization and separation of viral DNA polymerase in mouse milk. Proc. Natl Acad Sci USA 70:2117–2121

Kennedy N, Knedlitschek G, Groner B, Hynes NE, Herrlich P, Michalides R, van Ooyen AJJ (1982) Long terminal repeats of endogenous mouse mammary tumour virus contain a long open reading frame which extends into adjacent sequences. Nature 295:622–624

Kimball PC, Michalides R, Colcher D, Schlom J (1976) Characterization of mouse mammary tumor viruses from primary tumor cell cultures I. Immunogenic and structural studies. JNCI 56:111–117

Kohno M, Ishihama A (1979) Purification and properties of RNA-dependent DNA polymerase from cytoplasmic A-type particles of murine mammary tumor virus. Eur J Biochem 97:257–266

Lasfargues EY, Kramasky B, Sarkar NH, Lasfargues JC, Moore DH (1972) An established R111 mouse mammary tumor cell line: kinetics of mammary tumor virus production. Pro Soc Exp Biol Med 139:242–247

Lasfargues EY, Kramarsky B, Lasfargues JC, Moore DH (1974) Detection of mouse mammary tumor virus in cat kidney cells infected with purified B particles from milk. JNCI 53:1831–1833

Lasfargues EY, Lasfargues JC, Dion AS, Greene AE, Moore DH (1976) Experimental infection of a cat kidney cell line with the mouse mammary tumor virus. Cancer Res 36:67–72

Leavitt R, Schlesinger S, Kornfield S (1977) Tunicamycin inhibits glycosylation and multiplication of Sindbis and vesicular stomatitis viruses. J Virol 21:375–385

Lenz J, Crowther R, Straceski A, Haseltine W (1982) Nucleotide sequence of the Akv *env* gene. J Virol 42:519–529

Majors JE, Varmus HE (1981) Nucleotide sequences at host-proviral junctions for mouse mammary tumour virus. Nature 289:253–258

Marcus SL, Sarkar NH, Modak MJ (1976) Purification and properties of murine mammary tumor virus DNA polymerase. Virology 71:242–254

Marcus SL, Smith SW, Racevskis J, Sarkar NH (1978) The relative hydrophobicity of oncornaviral structural proteins. Virology 86:398–412

Marcus SL, Kopelman R, Sarkar NH (1979) Simultaneous purification of murine mammary tumor virus structural proteins: analysis of antigenic reactivities of native gp34 by radioimmunoprecipitation assays. J Virol 31:341–349

Massey RJ, Schochetman G (1979) Gene order of mouse mammary tumor virus precursor polyproteins and their interaction leading to the formation of a virus. Virology 99:358–371

Massey RJ, Arthur LO, Nowinski RC, Schochetman G (1980) Monoclonal antibodies identify individual determinants on mouse mammary tumor virus glycoprotein gp52 with group, class or type specificity. J Virol 34:635–643

Michalides R, van Nie R, Nusse R, Hynes NE, Groner B (1981) Mammary tumor induction loci in GR and DBAf mice contain one provirus of the mouse mammary tumor virus. Cell 23:165–173

Michalides R, Wagnenaar E, Hilkens J, Hilgers J, Groner B, Hynes NE (1982) Acquisition of proviral DNA of mouse mammary tumor virus in thymic leukemias from GR mice. J Virol 43:819–829

Moore DH, Long CA, Vaidya AA, Sheffield JB, Dion AS, Lasfargues EY (1979) Mammary tumor viruses. Adv Cancer Res 29:347–418

Nandi S, McGrath CM (1973) Mammary neoplasia in mice. In: Klein G, Weinhouse W (eds) Adv Cancer Res 17:353–414

Noon MC, Wolford RC, Parks WP (1975) Expression of mouse mammary tumor virus polypeptides in milks and tissues. J Immunol 115:653–658

Nowinski RC, Sarkar NH, Old LJ, Moore DH, Scheer DI, Hilgers J (1971) Characteristics of the structural components of the mouse mammary tumor virus II. Viral proteins and antigens. Virology 46:21–38

Nusse R, Asselbergs FAM, Solden MHL, Michalides RJAM, Bloemendal WH (1978) Translation of mouse mammary tumor virus RNA: precursor polypeptides are phosphorylated during processing. Virology 92:106–115

Nusse R, van der Ploeg L, van Duijn L, Michalides R, Hilgers J (1979) Impaired maturation of mouse mammary tumor virus precursor polypeptides in lymphoid leukemia cells producing intracytoplasmic A particles and no extracellular B-type virions. J Virology 32:251–258

Nusse R, Janssen H, de Vries L, Michalides R (1980) Analysis of secondary modifications of mouse mammary tumor virus proteins by two-dimensional gel electrophoresis. J Virol 35:340–348

Owens RB, Hackett AJ (1972) Tissue culture studies of mouse mammary tumor cells and associated viruses. JNCI 49:1321–1332

Panet A, Baltimore D, Hanafusa H (1975) Quantitation of avian RNA tumor virus reverse transcriptase by radioimmunoassay. J Virol 16:146–152

Parks WP, Howk RS, Scolnick EM, Oroszlan S, Gilden RV (1974) Immunochemical characterization of two major polypeptides from murine mammary tumor virus. J Virol 13:1200–1210

Peters G, Glover C (1980) tRNAs and priming of RNA-directed DNA synthesis in mouse mammary tumor viruses. J Virol 35:31–40

Peters G, Smith R, Brookes S, Dickson C (1982) Conservation of protein coding potential in the long terminal repeats of exogenous and endogenous mouse mammary tumor viruses. J Virol 42:880–888

Pless DD, Lennarz WJ (1977) Enzymatic conversion of proteins to glycoproteins. Proc Natl Acad Sci USA 74:134–138

Porter AG, Barber C, Carey NH, Hallewell RA, Threlfall G, Emtage JS (1979) Complete nucleotide sequence of an influenza virus hemagglutinin gene from cloned DNA. Nature 282:471–477

Racevskis J, Sarkar NH (1978) Synthesis and processing of precursor polypeptides to murine mammary tumor virus structural proteins. J Virol 25:374–383

Racevskis J, Sarkar NH (1979) Phosphorylation of murine mammary tumor virus precursor polypeptides. J Virol 30:241–247

Racevskis J, Sarkar NH (1980) Murine mammary tumor virus structural protein interactions: formation of oligomeric complexes with cleavable cross-linking agents. J Virol 35:937–948

Racevskis J, Sarkar NH (1982) ML antigen of DBA/2 mouse leukemias: expression of an endogenous murine mammary tumor virus. J Virol 42:804–813

Racevskis J, Karande KA, Sarkar NH (1981) Precursor product relationship between intracytoplasmic A particle and murine mammary tumor virus core proteins established by tryptic peptide analysis. Virology 109:201–205

Redmond SMS, Dickson C (1983) Sequence and expression of the mouse mammary tumour virus env gene. EMBO J 2:125–131

Ringold GM, Lasfargues EY, Bishop JM, Varmus HE (1975) Production of mouse mammary tumor virus by cultured cells in the absence and presence of hormones: assay by molecular hybridization. Virology 65:135–147

Ritzi E, Baldi A, Spiegelman S (1976) The purification of a gs antigen of the murine mammary tumor virus and its quantitation by radioimmunoassay. Virology 75:188–197

Robertson DL, Varmus HE (1981) Dexamethasone induction of the intracellular RNAs of mouse mammary tumor virus. J Virol 40:673–682

Sarkar NH, Dion AS (1975) Polypeptides of the mouse mammary tumor virus I. Characterization of two group-specific antigens. Virology 64:471–491

Sarkar NH, Whittington ES (1977) Identification of the structural proteins of the murine mammary tumor virus that are serologically related to the antigens of intracytoplasmic type A particles. Virology 81:91–106

Sarkar NH, Taraschi NE, Pomenti AA, Dion AS (1976) Polypeptides of the mouse mammary tumor virus II. Identification of two major glycoproteins with the viral structure. Virology 69:677–690

Sarkar NH, Whittington ES, Racevskis J, Marcus SL (1978) Phosphoproteins of the murine mammary tumor virus. Virology 91:407–422

Schloemer RH, Schlom J, Schochetman G, Kimball P, Wagner RR (1976) Sialylation of glycoproteins of murine mammary tumor virus, murine leukemia virus, and Mason-Pfizer monkey virus. J Virol 18:804–808

Schlom J, Hand PH, Teramoto YA, Callahan R, Todaro G, Schidlovsky G (1978) Characterization of a new virus from Mus cervicolor immunologically related to the mouse mammary tumor virus. JNCI 61:1509–1519

Schochetman G, Oroszlan S, Arthur LO, Fine DL (1977) Gene order of the mouse mammary tumor virus glycoproteins. Virology 83:72–83

Schochetman G, Long CW, Oroszlan S, Arthur LO and Fine DL (1978) Isolation of separate precursor polypeptides for the mouse mammary tumor virus glycoproteins and non-glycoproteins. Virology 85:168–174

Schwartz D, Tizard R, Gilbert W (1982) Appendix E, Complete nucleotide sequences of three retroviral genomes and a cellular onc gene. In: Weiss R, Teich N, Varmus H, Coffin J (eds) Molecular biology of tumor viruses. Part III, RNA tumor viruses, 2nd edn. Cold Spring Harbor Press, New York, pp 1338–1348

Sen A, Todaro GJ (1977) The genome-associated, specific RNA binding proteins of avian and mammalian type C viruses. Cell 10:91–99

Sen A, Sherr CJ, Todaro GJ (1977) Phosphorylation of murine type C viral p12 proteins regulates their extent of binding to the homologous RNA. Cell 10:489–496

Sen GC, Smith SW, Marcus SL, Sarkar NH (1979) Identification of the messenger RNAs coding for the *gag* and *env* gene products of the murine mammary tumor virus. Proc Natl Acad Sci USA 76:1736–1740

Sen GC, Haspel HC, Sarkar NH (1980a) Presence of a proteolytic activity in murine mammary tumor virus. J Biol Chem. 255:7098–7101

Sen GC, Zablocki W, Sarkar NH (1980b) Gene order of murine mammary tumor virus *gag* proteins and *env* proteins. Virology 106:152–154

Sen GC, Racevskis J, Sarkar NH (1981) Synthesis of murine mammary tumor viral proteins in vitro. J Virol 37:963–975

Sheffield JB, Daly TM (1976) Extrinsic labeling of MuMTV with a galactose oxidase-tritiated borohydride method. Virology 70:247–250

Sheffield JB, Daly T, Dion AS, Tarashi N (1977) Procedures of radioimmunoassay of the mouse mammary tumor virus. Cancer Res 37:1480–1485

Shinnick TM, Lerner RA, Sutcliffe JG (1981) Nucleotide sequence of Moloney murine leukaemia virus. Nature 293:543–548

Smith GH (1978) Evidence for a precursor-product relationship between intracytoplasmic A particles and mouse mammary tumor virus cores. J Gen Virol 41:193–200

Smith GH, Lee BK (1975) Mouse mammary tumour virus polypeptide precursors in intracytoplasmic A particles. JNCI 55:493–496

Smith GH, Wivel NA (1973) Intracytoplasmic A particles, mouse mammary tumor virus nucleoprotein cores? J Virol 11:575–584

Sykes JA, Whitescaver, Briggs L (1968) Observation on a cell line producing mammary tumor virus. JNCI 41:1315–1327

Takatushi A, Kohno K, Tamura G (1975) Inhibition of biosynthesis of polyisoprenol sugars in chick embryo microsomes by tunicamycin. Agricultural Biol Chem 39:2089–2091

Tanaka H (1977) Precursor-product relationship between non-glycosylated polypeptides of A and B particles purified from mouse tumors. Virology 76:35–350

Tanaka H, Moore DH (1967) Electron microscopic localization of viral antigens in mouse mammary tumors by ferritin-labeled antibody. Virology 33:197–214

Tanaka H, Tamura A, Tsujimura D (1972) Properties of intracytoplasmic A particles purified from mouse tumors. Virology 49:61–77

Tarentino AL, Maley F (1974) Purification and properties of an endo-$\beta$-$N$-acetylglucosaminidase from *Streptomycis griseus*. J Biol Chem. 249:811–817

Teramoto YA, Schlom J (1978) Radioimmunoassays demonstrating type-specific and group-specific antigenic reactivities for the major internal structural protein of murine mammary tumor viruses. Cancer Res 28:1990–1995

Teramoto YA, Schlom J (1979) Radioimmunoassays for the 36000-dalton glycoprotein of murine mammary tumor viruses demonstrate type, group, and interspecies determinants. J Virol 31:334–340

Teramoto YA, Puentes MJ, Young LJT, Cardiff RD (1974) Structure of the mouse mammary tumor virus: polypeptides and glycoproteins. J Virol 13:411–418

Teramoto YA, Cardiff RD, Lund JK (1977a) The structure of the mouse mammary tumor virus: isolation and characterization of the core. Virology 77:135–148

Teramoto YA, Kufe D, Schlom J (1977b) Type-specific antigenic determinants on the major external glycoprotein of high- and low-oncogenic murine mammary tumor viruses. J Virol 24:525–533

Teramoto YA, Hand PH, Callahan R, Schlom J (1980) Detection of novel murine mammary tumor viruses by interspecies immunoassays. JNCI 64:967–975

Vaidya AB, Lasfargues EY, Heubel G, Lasfargues JC, Moore DH (1976) Murine mammary tumor virus: characterization of infection of non-murine cells. J Virol 28:911–917

Vaidya AB, Long CA, Sheffield JB, Tamura A, Tanaka H (1980) Murine mammary tumor virus deficient in the major glycoprotein: biochemical and biological studies on virions produced by a lymphoma cell line. Virology 104:279–293

Varmus HE (1982) Form and function of retroviral proviruses. Science 216:812–820

Varmus HE, Ringold G, Yamamoto K (1979) Regulation of MMTV gene expression. In:

Baxter J, Rousseau G (eds) Glucocorticoid hormone action. Springer, Berlin Heidelberg New York, pp 254–278

Verstraeten AA, van Nie R, Kioa HG, Hageman PC (1975) Quantitative estimation of mouse mammary tumor virus (MMTV) antigens by radioimmunoassay. Int J Cancer 15:270–281

Weiss R, Teich N, Varmus H, Coffin J (eds) (1982) Molecular biology of tumor viruses. Part III, RNA tumor viruses. Cold Spring Harbor Press, New York

Wheeler DA, Butel JS, Medina DM, Cardiff RD, Hayes GL (1983) Transcription of mouse mammary tumor virus: identification of a candidate mRNA for the long terminal repeat gene product. J Virol 46:42–49

White J, Matlon K, Helenius A (1981) Cell fusion by Semliki Forest, influenza, and vesicular stomatitis viruses. J Cell Biol 89:674–679

Witte ON, Weissman IL, Kaplan HS (1973) Structural characteristics of some murine RNA tumor viruses studied by lactoperoxidase iodination. Proc Natl Aad Sci USA 70:36–40

Yagi MJ, Compans RW (1977) Structural components of mouse mammary tumor virus I. Polypeptides of the virion. Virology 76:761–766

Yagi MJ, Tomana M, Stutzman RE, Robertson BH, Compans RW (1978) Structural components of mouse mammary tumor virus III. Composition and tryptic peptides of virion polypeptides. Virology 91:291–304

Zangerle PF, Calberg-Bacq CM, Colin C, Franchemont P, Francois C, Gosselin L, Kozma S, Osterrieth PM (1977) Radioimmunoassay for glycoprotein gp47 of murine mammary tumor virus in organs and serum of mice and search for related antigens in human sera. Cancer Res 37:4326–4331

# Molecular Genetics of Mouse Mammary Tumor Virus

Vicki Traina-Dorge[1] and J. Craig Cohen[2]

# 1 Introduction

Infecting inbred mice with mouse mammary tumor virus (MMTV) provides a means for studying the interplay of genetic and epigenetic factors in oncogenesis. BITTNER (1936) showed that an extrachromosomal agent influenced the incidence of breast carcinoma in inbred mouse strains and thereby confirmed the horizontal transmission of the virus. Later work (for reviews

---

1 Department of Microbiology, Tulane University Medical Center, New Orleans, LA 70112, USA

2 Departments of Medicine and Biochemistry, Louisiana State University Medical Center, New Orleans, LA 70112, USA

Current Topics in Microbiology and Immunology, Vol. 106
© Springer-Verlag Berlin·Heidelberg 1983

see NANDI and McGRATH 1973; BENTVELZEN 1974) showed that genetic loci participate in mammary tumorigenesis, and several of these loci were shown to be endogenous viruses (VARMUS et al. 1972), indicating that vertical transmission of the virus occurred as well.

In most cells, infection and replication of MMTV does not interfere with cellular function or affect the cell's structure. Thus, the study of MMTV can provide information on the replication and synthesis of eukaryotic genes and their products. The hormone responsiveness of MMTV-specific genes (discussed by RINGOLD, p. 79) is of particular interest. However, in the mammary gland, cells become transformed after infection and adenocarcinomas develop; a classic case of virus-induced carcinogenesis and the only animal model for viral-induced malignancies of the breast. A third consequence of infection by this and other retroviruses is the acquisition of new, virus-specific genetic loci that segregate among all progeny of affected individuals according to Mendelian rules of genetics. Viral sequences that stably recombine with the mouse genome are called proviruses. Although MMTV proviruses may influence the incidence of malignancy, most are cryptic, having no discernible effect. The host must in some way repress the expression of genetically transmitted, virus-specific sequences, and tumorigenesis may be contingent on their derepression. Therefore, MMTV infection of germline cells is useful as a model system for the study of eukaryotic gene regulation.

In this paper, studies relating to the identification, organization, and expression of genetically transmitted MMTV sequences will be discussed. The development of this system will be traced from its origins in mouse genetics to its uses in molecular biology.

## 1.1 Inbred Mouse Development

The current understanding of MMTV infection and tumorigenesis is linked to the development of the inbred mouse (*Mus musculus*). To fully appreciate the genetics of this virus one must understand the methods used during inbreeding and the genealogy of these strains of mice. Therefore, this discussion begins with a consideration of the history and methods of mouse genetics.

Inbred mouse strain development began in the early 1900s when the possible genetic basis for cancer was recognized. Starting with stocks of colored mice originated in China and maintained by mouse fanciers, several investigators – including Little, Strong, Bagg, Snell, and others (MORSE 1978) – produced inbred colonies by successive brother/sister matings. An important fact that must be considered here is that inbred mouse strains did not originate from truly feral animals. This fact indicates that a nonrandom genetic stock was used to generate the most commonly used mouse strains. Therefore, as shown in Fig. 1, even though the selective breeding history of two strains may appear completely unrelated, the finding of simi-

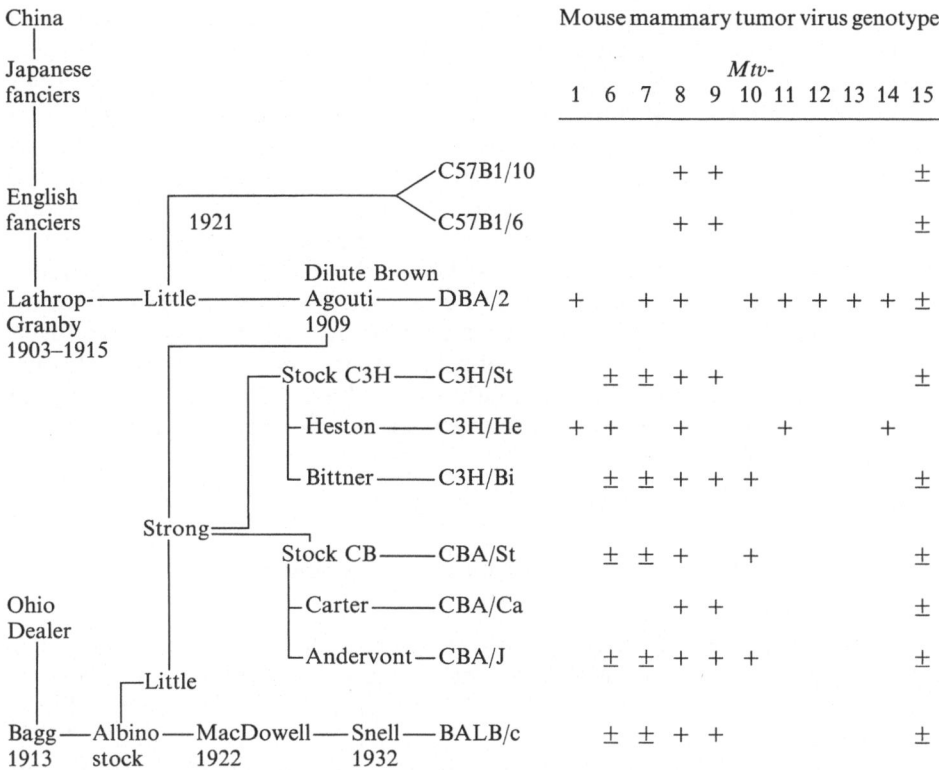

**Fig. 1.** The genealogy of the C3H/CBA family of inbred mice and the C57B1 and BALB/c strains (MORSE 1978). The diagram on the *right* demonstrates the polymorphic MMTV-specific alleles of each strain

lar viral loci reflects their presence in a common, pre-inbred ancestor obtained from the lay mouse fanciers.

### 1.1.1 Mouse Genetics

In mouse genetics, precise definitions and classifications were developed. An inbred strain is a colony of mice produced after a series of 20 or more brother/sister matings. This definition is based on the statistical inbreeding coefficient, which is a theoretical measure of inbreeding progress. Numerous substrains of any inbred line may occur due either to the distribution of an original stock before the 20th brother/sister mating or to the detection of genetic differences between colonies separated genetically for a long time. Examples of substrains are the C3H/He and C3H/Bi inbred mouse strains that were maintained separately by Heston and Bittner, respectively, and that were shown to have different MMTV proviruses (Fig. 1; COHEN and VARMUS 1979).

One can introduce genetic variation into inbred mouse strains by several methods. Obviously, crosses between two inbred lines may be used to generate genetically heterozygous mice. Congeneic inbred mice may occur either by a single mutation within an inbred strain or by crosses between two inbred mice and successive backcrosses to one of the parental strains to isolate a single allele from one parent in the genetic background of the other parent. Recombinant inbred (RI) strains are produced by crossing two established inbred mouse strains and random mating of the FI generation to permit reassortment of alleles within the FII progeny. Subsequent brother/sister mating from randomly paired FII parents results in fixation of parental polymorphic alleles. Thus, a series of inbred strains are produced that have segregated alleles polymorphic among the original parental strains. These strains are useful in linkage analyses and have permitted the mapping of MMTV proviruses and their assignment to specific chromosomes (TRAINA et al. 1981).

## 1.2 Mouse Mammary Tumor Virus DNA Synthesis

All retroviruses have a distinguishing characteristic: synthesis of a DNA intermediate from virion RNA by the virion-associated enzyme, reverse transcriptase (see WEISS et al. 1982). Unintegrated MMTV-specific DNA is isolated from infected heterologous cells (LASFARGUES et al. 1976; RINGOLD et al. 1977a, 1977b; SHANK et al. 1978; KUNG et al. 1981) and its synthesis indirectly stimulated by glucocorticoids (RINGOLD et al. 1978). From the cytoplasm of infected cells a linear molecule of about 9.8 kilobases (kb) is obtained. Nuclear preparations yield two covalently closed, super-coiled forms of MMTV-specific DNA having lengths of 9.8 and 8.6 kb.

Restriction endonuclease studies of these forms of viral DNA (SHANK et al. 1978) revealed that, like that of other retroviruses (see WEISS et al. 1982), MMTV DNA included two long terminal repeated (LTR) sequences which for MMTV were approximately 1200 basepairs (bp). Later, DNA sequencing of the MMTV LTR established their size at 1357 bp (DONE-HOWER et al. 1981). Examination of the relationship of the linear cytoplasmic viral DNA to nuclear forms indicated that the larger supercoiled molecule was formed by ligation of the ends of the linear DNA and the shorter by ligation of the linear with deletion of one copy of the LTR. An interesting feature of the MMTV LTR is the presence of an open reading frame that could code for a protein of about 36000 daltons (DONEHOWER et al. 1981; DICKSON et al. 1981; KENNEDY et al. 1982; PETERS et al. 1982). As discussed later, this may represent an as yet undefined MMTV gene.

Many strains or variants of MMTV can be distinguished by their tumorigenicity and immunogenic properties (see NANDI and MCGRATH 1973; BENTVELZEN 1974; ARTHUR et al. 1981). These viral strains are designated by several notations. In this review, the convention indicating the inbred mouse strain of origin as a postscript – e.g., MMTV(C3H) is MMTV from

C3H mice – is used. Although DNA from these strains exhibited greater than 95% sequence homology by solution hybridization kinetics (RINGOLD et al. 1976), restriction endonuclease digestions have revealed significant differences (COHEN et al. 1979a, b). Thus, the milk-borne MMTV(C3H) strain is distinguished from those transmitted genetically in either the BALB/c or C3H/He inbred mouse strains by a 4.2-kb fragment generated with *Pst*I. Likewise the milk-borne viruses of the C3H and GR mouse strains (MMTV(C3H) and MMTV(GR), respectively) are differentiated on the basis of a unique *Sac*I site (COHEN and VARMUS 1980).

Integration (genetic recombination) of MMTV DNA with host DNA is required for virus replication, as shown with all retroviruses (see WEISS et al. 1982). In both infected mouse mammary gland cells (COHEN et al. 1979b; HYNES et al. 1979) and in vitro infected heterologous cells (RINGOLD et al. 1979), viral DNA is integrated into a large number of, perhaps random, sites.

# 2 Genetic Transmission of Mouse Mammary Tumor Virus

In BITTNER's (1936) original studies the role of an extrachromosomal factor in mammary carcinogenesis was demonstrated in high-tumor strains of inbred mice. Thus, if the progeny from a C3H mouse strain having 90% tumor incidence was foster nursed at birth on a BALB/c female that normally has 1% tumor incidence, the resulting substrain, C3HfBALB/c (standard notation for a foster-nursed substrain) had a low tumor incidence, reflecting the preponderence of the milk-borne, virus-negative females. The reciprocal substrain, BALB/cfC3H, had a correspondingly high tumor incidence. However, both the C3HfBALB/c and BALB/c mice developed a low incidence of mammary tumors, indicating the involvement of genetically transmitted genes and leading to the suggestion that MMTV might be transmitted vertically (NANDI and MCGRATH 1973; BENTVELZEN 1974).

## 2.1 Distribution of Mouse Mammary Tumor Virus Sequences

Nucleic acid hybridization between MMTV-specific complementary DNA (cDNA) and host DNAs (VARMUS et al. 1972; PARKS and SCOLNICK 1973; GILLESPIE et al. 1973; MICHALIDES and SCHLOM 1975; MCGRATH et al. 1978) established the presence of genetically transmitted viral sequences in the genome of all inbred strains of mice. These data confirmed the conclusions based on the appearance of virus-positive tumors in animals free of milk-borne virus. In addition to finding these sequences in the mouse, MMTV-related sequences were detected in DNA samples from rats (DROHAN and

SCHLOM 1979 b) and human mammary tumors (DAS and MINK 1979; CALLA-
HAN et al. 1982).

MORRIS et al. (1977) quantitated the MMTV-specific sequences in DNA
from inbred and feral mice, as well as members of two species of Asian
mice (*Mus cervicolor* and *M. caroli*). In DNA from inbred mice, from five
(RIII strain) to ten (GR strain) copies/per cell of MMTV were found in
normal organs. Pooled samples from feral animals were shown to have
three copies per cell. Characterization of these sequences in DNA from
both *M. cervicolor* and *M. caroli* revealed sequences with only partial homology with MMTV (about 30% and 10%, respectively). However, in *M. cervicolor*, over 30 copies per cell of this MMTV-related sequence were detected.

Nucleic acid hybridization was used to compare the sequence homology
of different MMTV strains. Although studies of RNase Tl oligonucleotides
found differences among MMTV isolates (FRIEDRICH et al. 1976), RINGOLD
et al. (1976) demonstrated greater than 96% sequence homology among
MMTV strains. DROHAN and SCHLOM (1979 a) showed that sufficient sequence heterogeneity occurred between MMTV(C3H) and MMTV(C3Hf)
to permit selection of a MMTV(C3H)-specific hybridization probe from
viral RNA. As discussed later, restriction endonuclease mapping confirmed
the overall homology of these strains but with minor base changes (SHANK
et al. 1978; COHEN et al. 1979a; COHEN and VARMUS 1980; DROHAN et al.
1981).

## 2.2  Phenotypic Segregation Studies

The genetics of MMTV in the inbred mouse are reviewed extensively elsewhere (NANDI and MCGRATH 1973; BENTVELZEN 1974; BENTVELZEN et al.
1978). These studies resulted in the characterization of five MMTV-specific
loci (*Mtv-1* to *-5*; Table 1). This discussion is limited to those aspects ad-
dressed by studies of MMTV molecular biology.

The identification of MMTV-specific genetic loci and chromosomal location were determined initially using crosses between high and low expressor
strains. Before the development of techniques for the direct visualization
of individual MMTV proviruses, expression of the locus resulting in either
increased tumorigenicity or synthesis of virus-specific proteins was required.
In the FI cross, dominant genes were uniformly expressed in all progeny
but recessive genes are not. Backcrosses between the FI generation and
the nonexpressor strain of a dominant gene result in about 50% expressors.
Similarly, a backcross between the FI and the expressor strain of a recessive
gene produced 50% expressors. In addition to determining the dominant
or recessive nature of a gene, these methods revealed the chromosomal
location. Co-segregation during the backcrosses of the virus-specific locus
with any one of the numerous nonviral loci mapped on the mouse genome
(O'BRIEN 1980) indicated linkage on the same chromosome.

**Table 1.** Molecular characterization and chromosomal location of MMTV-specific genetic loci

| Locus | EcoRI fragment(s) (kb) | Chromosome location | Reference |
|---|---|---|---|
| Mtv-1 | 6.5, 4.5 | 7 | 1–5 |
| Mtv-2 | 11.7, 6.9 | 18 | 5–9 |
| Mtv-3 | ND | 11 | 7, 10 |
| Mtv-4 | ND | ? | 14 |
| Mtv-5 | ND | ? | 14 |
| Mtv-6 | 16.7 | ? | 2, 11 |
| Mtv-7 | 16.7 | 1 | 4 |
| Mtv-8 | 8.5, 6.7 | ? | 2, 4, 11, 13 |
| Mtv-9 | 10.0, 7.8 | ? | 2, 4, 11, 13 |
| Mtv-10 | 11.7, 10.0 | 1 | 2, 4 |
| Mtv-11 | 5.8 | ? | 2, 4 |
| Mtv-12 | 15.0 | ? | 2, 4 |
| Mtv-13 | 5.8, 9.0 | ? | This paper |
| Mtv-14 | 1.7 | 6 | 4, 13 |
| Mtv-15 | 10.0 | ? | 4 |
| Mtv-16 | 7.8, 6.4 | ? | 12 |
| Mtv-17 | 13.0, 5.3 | ? | 15 |

(1) VAN NIE and VERSTRAETEN 1975, (2) COHEN and VARMUS 1979, (3) COHEN and VARMUS 1980, (4) TRAINA et al. 1981, (5) MICHALIDES et al. 1981, (6) VAN NIE et al. 1977, (7) VAN NIE and DEMOES 1977, (8) MICHALIDES et al. 1978, (9) FANNING et al. 1980, (10) NUSSE et al. 1980, (11) COHEN et al. 1979b, (12) HYNES et al. 1981, (13) LONG et al. 1980, (14) HILGERS, personal communication, (15) VAIDYA, personal communication

## 2.2.1 Mtv-1

In C3H and related inbred mouse strains (Fig. 1) it was noted that if C3Hf substrains were foster nursed on virus-free females the incidence of mammary tumors was significantly reduced (22%–55%) (HESTON and DERINGER 1952; BOOT and MUHLBOCK 1956; HESTON 1958); however fosternursing did not lower the incidence to that of other virus-free strains, e.g., BALB/c, 1% (NANDI and McGRATH 1973). Electron-microscopic examination of both normal and neoplastic tissues of the C3Hf strain revealed virus particles typical of MMTV (BERNHARD et al. 1956; PITELKA et al. 1960). This virus, MMTV(C3Hf), was transmitted to other strains (e.g., BALB/c) in which it could be maintained as a milk-borne virus. However, this virus was not as oncogenic as the MMTV(C3H) virus.

The genetic loci required for MMTV(C3Hf) production was characterized by VAN NIE and VERSTRAETEN (1975); VERSTRAETEN and VAN NIE (1978). Using crosses between either the C3Hf or DBAf and BALB/c mouse strains, the dominance of this locus was demonstrated. Backcrosses of the F I to the nonexpressor parent (BALB/c) resulted in about 50% expressors, consistent with a dominant genetic locus. This gene was designated Mtv-1, and, because it segregated with the albino locus, it was mapped to chromosome 7 in both C3H and DBA strains.

### 2.2.2 *Mtv-2*

The European mouse strain, GR, was shown to exhibit a high tumor incidence (90%) in the absence of a milk-borne virus. Transmission occurred
with either males or females, indicating an inheritable trait (VAN NIE et al.
1977). Data from crosses between GR and low-tumor-incidence strains and
subsequent backcrosses (MICHALIDES et al. 1978) indicated that a single
dominant gene was required. Linkage between this locus, *Mtv-2*, and the
*Tw* (twirler) locus on chromosome 18 was demonstrated. However, unlike
*Mtv-1* which encoded a virus of low oncogenicity expressed infrequently,
*Mtv-2* viral sequences are   highly oncogenic and uniformly expressed in
all GR females.

### 2.2.3 *Mtv-3*

A congeneic GR substrain was developed that was negative for the *Mtv-2*
locus (GR/*Mtv-2-*; VAN NIE and DEMOES, 1977; NUSSE et al. 1980). In this
strain, production of the MMTV-specific core protein (*gag* gene) − but
not the envelope glycoprotein (*env* gene) − was detected by radioimmunoassay. Genetic analysis of this locus indicated the involvement of a single
dominant gene, *Mtv-3*, located on chromosome 11. This locus was unique
because only partial expression of the viral genome occurred.

## 3   Molecular Characterization of Mouse Mammary Tumor Virus-Specific Loci

The solution hybridization and genetic studies reveal much concerning the
genetic basis of virus-induced mammary carcinogenesis in the mouse. However, neither of these methods can be used to assess the organization of
MMTV-specific DNA transmitted genetically in the mouse. Thus, describing
the size or the genetic content of individual proviruses is impossible. Furthermore, these studies rely on pooled samples, which makes study of the individual in feral populations difficult.

### 3.1   Restriction Endonuclease Mapping

The introduction of molecular analysis with restriction endonucleases, involving gel electrophoresis and DNA transfer to nitrocellulose for the detection of MMTV-specific sequences (SOUTHERN 1975), has greatly advanced
the understanding of the genetically transmitted retroviruses. The restriction
endonucleases, *Pst*I and *Bam*HI, which cleave at several sites within MMTV

DNA (SHANK et al. 1978), and *Eco*RI, which cleaves at a single site, have proved to be useful in the characterization of MMTV proviruses.

MMTV sequences acquired by milk-borne infection are distinguished from many of those endogenous to various inbred strains of mice by slight but distinct differences in internal *Pst*I restriction cleavage sites. Digestion with *Pst*I of DNA from MMTV(C3H) transmitted in the milk yields unique fragments of 4.2 and 1.0 kb (COHEN et al. 1979b). However, in some inbred mouse strains, e.g., C3H/Bi (Fig. 1; COHEN and VARMUS 1979), an endogenous provirus that yields *Pst*I fragments identical to MMTV(C3H) is present. To distinguish acquired from endogenous viral sequences, therefore, one must ultimately rely on restriction endonucleases that cleave at a single site in the viral genome, e.g., *Eco*RI (SHANK et al. 1978). Because retroviruses generally integrate at many, perhaps random, sites in the host genome, each infection and integration event generates two unique fragments after digestion with an enzyme with only a single cleavage site in viral sequences (BOTHCHAN et al. 1976). This characteristic of virus replication is used to distinguish both acquired and endogenous proviruses occurring at unique sites in the mouse genome.

## 3.2 Molecular Segregation Studies

In most studies, identification of individual MMTV-specific proviruses depended largely on their segregation among inbred strains (COHEN and VARMUS 1979; HYNES et al. 1979; GRONER and HYNES 1980; LONG et al. 1980; COHEN and VARMUS 1980). However, isolation of individual fragments generated by digestion with *Eco*RI followed by extensive restriction endonuclease mapping was used to characterize the proviruses endogenous to the BALB/c mouse strain (COHEN et al. 1979a). Three MMTV-specific proviruses were demonstrated. Two of these proviruses were complete, i.e., about 10 kb in size with LTRs, but one was subgenomic, having only 2- to 3-kb of viral sequences. The use of molecular cloning in bacteriophage and plasmid vectors confirmed and extended these findings (GRONER et al. 1980; DONEHOWER et al. 1981; HYNES et al. 1981a; HERRLICH et al. 1981; BUETTI and DIGGELMANN 1981).

### 3.2.1 Inbred Mice

Although most endogenous MMTV proviruses are phenotypically silent, these loci in many American inbred mouse colonies were characterized using restriction endonucleases (COHEN and VARMUS 1979; COHEN and VARMUS 1980; HYNES et al. 1979; GRONER and HYNES 1980; LONG et al. 1980). The C3H CBA family of inbred mice was derived by Strong from the Bagg albino (BALB/c) stock and the Little dilute brown agouti (DBA) stock (see Fig. 1). Digestion of DNA from the progeny-inbred strains with restriction endonucleases (*Eco*RI and *Pst*I) revealed polymorphic, MMTV-specific fragments (COHEN and VARMUS 1979; GRONER and HYNES 1980). Analysis of the polymorphic MMTV-specific fragments found in the parental strains and segregating among the progeny strains identified six virus-specific loci.

From the identification and segregation of MMTV loci in these strains, several conclusions were made concerning the distribution, stability, evolution, and role in oncogenesis of genetically transmitted MMTV.

1. Because loci in inbred strains, separated genetically for many years, e.g., DBA/2 and C3H/Bi (Fig. 1), remain unaltered, genetically transmitted MMTV sequences are stable.
2. The observation that all loci in the progeny can be traced to the parental strains suggests that acquisition of new MMTV-specific proviruses in the germline is a rare event. This rarity is particularly evident in widely divergent strains (e.g., C57BL/10 and BALB/c) that include similar proviruses, but with genetic contact limited to pre-inbred strain development (Fig. 1).
3. Because these viral sequences are located at multiple sites within the mouse genome, they are obviously acquired by independent infection events of germline cells.

### 3.2.2 Feral Mice

Characterization of MMTV-specific sequenes of feral mice (COHEN and VARMUS 1979; CALLAHAN et al. 1982) substantiated the above-listed conclusions. Virus-specific fragments varied widely among DNA samples derived from individual animals, as expected if these sequences were acquired by random, independent, and rare infections. Furthermore, MMTV-specific loci in the feral animals were not identical to those of the inbred mouse strains. Finally, mice completely devoid of all MMTV-specific sequences were found in both *M. musculus* (COHEN and VARMUS 1979; COHEN et al. 1982) and related species (CALLAHAN et al. 1982).

### 3.2.3 Recombinant Inbred and Congeneic Mice

Segregation analysis among inbred mice of MMTV-specific fragments proved to be valuable in defining viral genetic loci. However, the varieties of genetically related inbred mice are limited, thus restricting the number of polymorphic virus-specific alleles. Recombinant inbred and congeneic inbred mice have extended the value of segregation analysis.

Recombinant inbred strains derived from the low-mammary-tumor-incidence strain, C57BL/6, and either DBA/2 or C3H/HeJ (B × D and B × H strains, respectively), which are high in tumor incidence, were used (TRAINA et al. 1981). The 23 B × D and 11 B × H RI strains are greater than 99% homozygous for all loci but have segregated those alleles polymorphic in the parental strains (see Fig. 1). Segregation of *Eco*RI restriction fragments among these RI strains was used to confirm earlier defined loci and to extend molecular characterization to 11 separate virus-specific loci in these mice (Table 1).

Among the MMTV-specific genetic loci it was noted that several (*Mtv-6, -7, -11, -12, -14, -15*) lacked an *Eco*RI cleavage site, indicating that they were either mutated at this site or are subgenomic, as shown with the *Mtv-6* locus in BALB/c (COHEN and VARMUS 1980). Because this site is uniform in DNA from all MMTV strains mapped (SHANK et al. 1978; COHEN et al. 1979a; GRONER and HYNES 1980; COHEN and VARMUS 1980), those proviruses lacking a *Eco*RI site are considered subgenomic. The remaining proviruses included an MMTV genome equivalent. The replicative process involved in insertion of these aberrant, genetically transmitted proviruses could require (a) an error-prone DNA synthesis and integration process that destroys this site at a  high frequency, (b) deletion of sequences that include the *Eco*RI site from the provirus, or (c) reverse transcription of mRNA species, resulting in subgenomic proviruses. MAJORS and VARMUS (1981) demonstrated that the envelope mRNA could serve as template for synthesis of a provirus, supporting the last hypothesis.

A problem with segregation analysis of retrovirus-specific fragments in inbred mice is the difficulty encountered with the comigration of identical, or nearly identically sized, fragments. In the C3H/CBA family of inbred mouse strains (TRAINA et al. 1981), three MMTV genetic loci include a 10.0-kb fragment. Two of these proviruses are genome length and one is presumed subgenomic. Similarly, the 5.8-kb fragment may derive from multiple proviruses, because its distribution is not random (23 of 26 positive strains) among the RI strains (Table 2; $P < 0.001$). In statistically comparing the segregation of the 5.8, 15.0, and 9.0 kb MMTV-specific fragments in the RI strains, we noted that with 14 of 26 samples the 5.8-kb fragment is present whenever either the 15.0- or 9.0-kb fragments are present whenever either the 15.0- or 9.0-kb fragments are present (Table 2). When the significance is calculated using Fisher's Exact test for the association of the 5.8-kb fragment with either the 9.0- or the 15.0-kb *Eco*RI fragments, a highly significant correlation ($P < 0.01$) is obtained. However, independent testing reveals only the association of the 5.8-kb fragment with the 9.0-kb fragment ($P < 0.02$), but not the 5.8- with the 15.0-kb fragment. In the B × H strains, the 5.8-kb fragment is present, but both the 15.0- and 9.0-kb fragments are absent. In these two strains the fragment exhibits random segregation and thus only a single, unlinked copy occurs. Although further segregation analysis is required, one can conclude that the 5.8-kb fragment may derive from two MMTV-specific loci (Table 1), *Mtv-11* (5.8 kb) and *Mtv-13* (9.0 and 5.8 kb).

DNA from the GR mouse strain, originated in Europe, were examined for MMTV-specific sequences after restriction endonuclease digestion (SHANK et al. 1978; MICHALIDES et al. 1978; GRONER and HYNES 1980; HYNES et al. 1981a; MICHALIDES et al. 1981b; ETKIND et al. 1982). Viral loci in this and other European inbred mouse strains, RIII and 020 (GRONER and HYNES 1980; LONG et al. 1980; ETKIND et al. 1982), as well as Asian

**Table 2.** Distribution of mouse mammary tumor virus-specific *Eco*RI fragments in B × D RI mouse strains[a]

| *Eco*RI fragment (kb) | Parental strain[b] | | B × D mouse strain | | | | | | | | | | | | | | | | | | | | | | | | | | |
|---|---|---|---|---|---|---|---|---|---|---|---|---|---|---|---|---|---|---|---|---|---|---|---|---|---|---|---|---|---|
| | B | D | 1 | 2 | 5 | 6 | 8 | 9 | 11 | 12 | 13 | 14 | 15 | 16 | 18 | 19 | 20 | 21 | 22 | 23 | 24 | 25 | 27 | 28 | 29 | 30 | 31 | 32 |
| 15.0 | − | + | − | + | − | + | + | + | + | − | − | + | + | + | − | + | − | − | + | − | + | + | − | − | + | − | + | + |
| 5.8 | − | + | + | + | + | + | + | + | + | + | − | + | + | + | + | + | − | + | + | + | − | + | + | + | + | + | + | + |
| 9.0 | − | + | + | − | + | − | − | + | − | + | − | + | + | + | + | + | − | + | + | + | − | + | + | + | + | + | − | + |

[a] From Traina et al. (1981)
[b] B, C57B1/6J; D, DBA/2J

inbred mouse strains, FM (HYNES et al. 1979), appeared unrelated to those present in mice bred in the United States.

These data support the concept of random independent infection of germline sequences as the origin of genetically transmitted retroviruses. However, some differences in the MMTV-specific fragments were reported for DNA from the same inbred strain but from different laboratories. *Eco*RI fragments generated in the RIII strain vary from five (ranging from 10.0 to 1.7 kb) (LONG et al. 1980) to six (ranging from 18.0 to 6.7 kb) (HYNES et al. 1979). Differences were also reported for DNA from the GR and A inbred mouse strains (MORRIS et al. 1979; LONG et al. 1980; GRONER and HYNES 1980; COHEN and VARMUS 1980). The maps of restriction sites within the MMTV(GR) and MMTV(C3H) viruses exhibited only minor differences. However, MMTV(RIII) (ETKIND et al. 1982) differs greatly from either of these two viruses, suggesting a major evolutionary shift in this viral strain.

## 3.3 Chromosomal Location

Although restriction endonuclease digestion studies indicated that MMTV genetic loci occurred at many sites in the mouse genome, these data cannot distinguish between random integration on one or multiple chromosomes. As shown with baboon endogenous virus (LEMONS et al. 1977, 1978; COHEN and MURPHEY-CORB 1983), a retrovirus may require a host locus for integration and replication. Therefore, the chromosome distribution of these viruses is important.

The genetic linkage studies of the phenotypically expressed loci, *Mtv-1*, *-2*, and *-3* (see above), indicate the involvement of multiple chromosomes, because these loci were mapped to chromosomes 7, 18, and 11, respectively. Using mouse × hamster hybrid cells that segregate mouse chromosomes, MORRIS et al. (1979) demonstrated that MMTV-specific loci in the A/HeJ mouse strain were located on multiple chromosomes. One provirus was mapped to chromosome 4, and other virus-specific alleles appeared to segregate with at least three separate chromosomes.

By combining genetic linkage and restriction endonuclease analyses, determining the chromosomal location of several viral alleles in RI mouse strains became possible (TRAINA et al. 1981). *Eph-1*, *Mtv-7*, *Mls*, *Mtv-10*, and *Ly-9* genetic markers and MTV-specific alleles co-segregated, indicating their location on chromosome 1, and mapped in the presented order on that chromosome. On chromosome 6, *Mtv-14*, *Lyt-2*, and *Ggc* were localized in the given orientation. Of particular interest was the finding (MICHALIDES et al. 1981 b; TRAINA et al. 1981) that one MMTV-specific allele segregated with the *Gpi-1*, *Tam-1*, *Mdr*, and *Hbb* alleles on chromosome 7 at a position between the *Tam-1* and *Mdr* loci. This position is identical to that of the *Mtv-1* locus mapped by VAN NIE and VERSTRAETEN (1975, 1978) in the C3Hf and DBAf mouse strains. Therefore, for the first time one can conclude that a locus defined genetically by means of phenotype segregation and a molecularly characterized locus (6.5 and 4.5-kb *Eco*RI fragments) are the same.

MICHALIDES et al. (1981 a) analyzed congeneic mouse strains that were either negative for the *Mtv-2* locus (GR/*Mtv-2*-; VAN NIE and DEMOES 1977) or positive for *Mtv-2* on a heterologous strain background (020/*Mtv-2*+). In both congeneic mice, 11.0- and 6.9-kb *Eco*RI fragments were shown to segregate with the *Mtv-2* phenotype.

# 4 Tumorigenesis and Endogenous Mouse Mammary Tumor Virus

## 4.1 Milk-Borne-Virus-Induced Tumors

Mammary tumors may develop from infection with either the milk-borne (BITTNER 1936) or the genetically transmitted MMTV (VAN NIE and VER-STRAETEN 1975, 1978). Characterization of virus-specific sequences in DNA from tumors induced by exogenous virus infection in several mouse strains (COHEN et al. 1979b; HYNES et al. 1979; GRONER and HYNES 1980; COHEN and VARMUS, 1980; FANNING et al. 1980a, b) indicates that tumors are largely clonal. The clonality of mammary tumors suggests that only one, or a few cells, of the many infected cells in the mammary gland undergoes malignant transformation. Recently, NUSSE and VARMUS (1982) found in many tumors that MMTV proviruses are associated with a specific region of the mouse genome. Although this region seems large (20 kb), this finding supports the concept of a promoter insertion mechanism of mammary tumorigenesis, similar to that proposed for leukosis viruses (HAYWARD et al. 1981; VARMUS et al. 1981; PAYNE et al. 1982). This model proposes that retroviruses alter the expression of cellular oncogenes (genes required for initiation and maintenance of the transformed state) via integration into sequences near or immediately adjacent.

## 4.2 Endogenous Virus-Induced Tumors

In tumor DNA from mice that are not infected by the milk-borne virus, viral DNA replication and integration does not always occur. In spontaneous tumors from the BALB/c mouse, no new MMTV-specific sequences were found (BREZNIK and COHEN 1982), indicating that, although virus replication and integration greatly facilitate tumorigenesis (99% tumors in the BALB/cfC3H substrain compared with 1% in BALB/c), neither is required. Furthermore, in GR/N mice many tumors were also found that do not show evidence of virus integration (BREZNIK and COHEN, unpublished observation).

ETKIND and SARKAR (1983) examined tumor DNA from C3Hf mice, whose tumor incidence of 22%–55% is attributed to the *Mtv-1* locus (VAN NIE and VERSTRAETEN 1975, VERSTRAETEN and VAN NIE 1978; Table 1; Fig. 1). In these tumors, amplification of the endogenous provirus was observed. Thus, expression of the *Mtv-1* locus and subsequent viral DNA replication and integration facilitated tumorigenesis, explaining the relatively high tumor incidence in the C3Hf mice compared with that of the BALB/c strain.

## 4.3 Regulation of Endogenous Mouse Mammary Tumor Virus Expression

Of the more than 15 MMTV-specific alleles described in inbred mice (Table 1), fewer than half show any detectable expression. Therefore, in the eukaryotic cell, active suppression of these viral genes must occur and the reversal of repression might be an essential step in tumorigenesis.

### 4.3.1 Mouse Mammary Tumor Virus DNA Methylation

In cells of eukaryotes, the predominant modified base is 5-methylcytosine (5mC) (for reviews see RAZIN and RIGGS 1980; EHRLICH and WANG 1981), unlike prokaryotes, which have 6-methyl adenosine in addition to 5mC. In prokaryotes, modified bases are thought to be involved in host restriction; however, their role in eukaryotes is not completely delineated. Several investigators (RIGGS 1975; HOLLIDAY and PUGH 1975) have implicated methylation in gene repression.

Using restriction endonucleases that differentiate methylated from unmethylated DNA sequences, e.g., *Msp*I and *Hpa*II (BIRD and SOUTHERN 1978), the MMTV-specific sequences of normal and neoplastic tissues were analyzed (COHEN 1980). Proviruses acquired by milk-borne infection (BALB/cfC3H substrain) were shown to be hypomethylated, However, those viral sequences transmitted genetically (*Mtv-6, -8,* and *-9*) were hypermethylated. Because the acquired proviruses are expressed and the endogenous are not, an inverse correlation between gene expression and methylation was proposed. A similar relationship was observed with other viral genes (DESROSIERS et al. 1979; SUTTER and DOERFFER 1980) and confirmed in other studies of MMTV (COHEN, 1980; HYNES et al. 1981b; FANNING et al. 1982) and other nonviral genes (MCGHEE and GINDER, 1979; MANDEL and CHAMBON 1979).

The role of DNA methylation in gene regulation in eukaryotes suggests a possible mechanism for the activation of genetically transmitted retroviruses. Because these virus-specific alleles are highly methylated and repressed, demethylation of these sequences could either initiate their expression or at least provide a useful marker for expression.

### 4.3.2 Mouse Mammary Tumor Virus DNA Demethylation

When MMTV proviruses in DNA from BALB/c tumors were examined for methylation, it was found that specific demethylation of *Mtv-6* occurred (BREZNIK and COHEN 1982). Interestingly, *Mtv-6* is a subgenomic locus that includes between 2- to 3-kb viral sequences derived from the 3' end of virion RNA (COHEN et al. 1979a). In addition, in one tumor only the 3' end of *Mtv-8* and *-9*, which includes complete proviruses, were demethylated, while the 5' half remained methylated. Characterization of DNA methylation in preneoplastic hyperplastic outgrowths (MEDINA 1973; CARDIFF et al. 1977; CARDIFF et al. 1981; BREZNIK, BUTEL, MEDINA, and COHEN, unpublished observation) indicated that the demethylation of *Mtv-6* was unique to the malignant cell. Furthermore, treatment of mice with transplanted hyperplastic outgrowth lines with 5-azacytidine, an inhibitor of methylation, increased the frequency of mammary tumors and certain demethylated MMTV-specific sequences.

ETKIND and SARKAR (1983) used DNA methylation to study tumors of the C3Hf strain. In DNA from these tumors, which exhibit amplification of the *Mtv-1* locus, the two *Eco*RI fragments that define this locus were demethylated. Therefore, altered methylation, expression, and viral DNA integration are evident with the *Mtv-1* locus. Similarly, in DNA from the GR mouse, which exhibits 100% tumor incidence in the absence of milk-borne virus, demethylation of MMTV-specific sequences was found in DNA from both mammary tumors and lactating mammary glands (BREZNIK and COHEN, unpublished observation).

Analysis of endogenous proviruses of the C3H/Bi strain revealed that the normally milk-borne, highly oncogenic MMTV(C3H) virus strain was transmitted genetically (*Mtv-10*) (Table 1; COHEN and VARMUS 1979; TRAINA et al. 1981). Digestion with methyl-sensitive restriction endonucleases indicated that this MMTV-specific genetic locus in the C3H/Bi mouse was hypomethylated in normal cells (COHEN 1980). Interestingly, this mouse strain was shown to have a high frequency of male transmission of mammary tumors to offspring of females from uninfected strains (HILGERS and BENTVELZEN 1978). Because hypomethylation correlates with expression, the finding that this locus is hypomethylated may explain this biological phenomenon.

The function of DNA methylation and demethylation in the initiation of mammary tumorigenesis is still not resolved. However, because the inverse correlation between methylation and gene expression is so strong, some conclusion can be drawn as to the possible mechanism of tumorigenesis by these genetically transmitted viral sequences. The demethylation and amplification of the *Mtv-1* locus clearly indicates that endogenous viruses can become activated from a dormant state and induce both virus production and malignant transformation.

The altered methylation of the subgenomic *Mtv-6* locus is enigmatic. If one accepts the notion that MMTV-induced oncogenesis involves the promoter insertion model (HAYWARD et al. 1981; PAYNE et al. 1982; NUSSE and VARMUS 1982), then one could envision that the MMTV promoter present in 3'-specific viral sequences becomes activated. This promoter activation model depends on the presence of an oncogene near the *Mtv-6* locus. In addition to activating cellular genes, promoter demethylation could activate an as yet unidentified viral gene. DNA sequencing of the MMTV LTR indicates the presence of an open reading frame that could encode a gene of about 36000 daltons (DONEHOWER et al. 1981; DICKSON et al. 1981). In the tissues from the BALB/c mouse, MMTV-specific RNA is detected but includes sequences from the 3' end of viral RNA. In addition, Hagar's group (personal communication) has identified a specific RNA species that could be transcribed from the *Mtv-6* locus. Therefore, the demethylation of this subgenomic viral sequence may indicate a second model for mammary tumorigenesis, one involving the protein transcribed from the open reading frame within the LTR.


# 5  Conclusions

In the mouse, MMTV-specific loci are acquired by apparently random, independent infection events of germline cells. These sequences are remarkably stable once recombination with host DNA occurs. Most virus-specific loci are phenotypically silent, having no apparent effect on the host. However, several loci are activated and result in mammary tumors.

The molecular mechanism for virus-induced mammary tumorigenesis in the mouse is still unclear. Data indicating some preferred MMTV DNA integration sites in DNA from mammary tumors suggest that promoter insertion and activation of host oncogenes may occur. This could certainly explain the synergistic effect of hormones and virus in oncogenesis, because the MMTV promoter is a glucocorticoid response; thus, the activated oncogene would now be stimulated by hormones. However, the finding of a virus-specific RNA transcribed from the open reading frame in the viral DNA LTR may be indicative of a virus-specific gene product required for tumorigenesis.

The studies summarized here reveal the value of MMTV infection in the mouse for examining tumorigenesis and gene expression in mammalian cells. Our understanding of the viral genetics is extended from the biological phenomena observed in various inbred populations to the molecular organization and location of individual virus-specific loci. However, several areas remain unclear but can be probed using currently available biological and technological resources.

First, although our knowledge of the molecular organization of MMTV-specific loci is the most extensive of any retrovirus system, further clarification and refinement of these data, including their chromosomal location and viral sequence complexity, must be obtained.

Second, differential expression of MMTV-specific loci, even though these viral sequences and integration sites in various inbred mice appear identical, is a continuing enigma. Are there nonviral loci which act in *trans* to either repress or enhance expression or are there *cis* elements which are polymorphic in the mouse?

Third, hormones and chemical carcinogens facilitate tumorigenesis, yet their relationship to endogenous viral gene expression is not defined. This is especially important in relating the mouse model to human disease. Because the familial pattern observed in humans suggests a genetically transmitted, possibly viral sequence which becomes activated, understanding the role of these various agents is essential if the experimental model is to have any relevance in medicine.

*Acknowledgements.* The author's laboratory is supported by USPHS grant number R01-CA-34823. VT-D is supported by a Hanna Horner fellowship. JCC is supported by the Optomist Leukemia Foundation of Louisiana, Inc.

# References

Arthur LO, Altrock BW, Schochetman G (1981) Type-specific determinants on proteins of an endogenous C3H mouse mammary tumor virus (MMTV) distinguish this virus from highly oncogenic exogenous MMTVs. Virology 110:270

Bentvelzen P (1974) Host-virus interactions in murine mammary carcinogenesis. Biochem Biophys Acta 355:236–259

Bentvelzen P, Brinkhof J, Haaijman JJ (1978) Genetic control of endogenous murine mammary tumour viruses reinvestigated. Eur J Cancer 14:1137–1147

Bernhard W (1958) Electron microscopy of tumor cells and tumor viruses a review. Cancer Research 18:494–509

Bird AP, Southern EM (1978) Use of restriction enzymes to study eucaryotic DNA methylation. I. The methylation pattern of ribosomal DNA from *Xenopus laevis*. J Mol Biol 118:27–47

Bittner JJ (1936) Some possible effects of nursing on the mammary gland tumor incidence in mice. Science 84:162–164

Boot LM, Muhlbock O (1956) The mammary tumor incidence in the C3H mouse strain with and without agent (C3H; C3Hf; C3He). Acta Univ Int Ca 12:569–581

Botchan M, Topp W, Sambrook J (1976) The arrangement of simian virus 40 sequences in the DNA of transformed cells. Cell 9:269–287

Breznik T, Cohen JC (1982) Altered methylation of endogenous viral promoter sequences during mammary carcinogenesis. Nature 295:255–257

Buetti E, Diggelmann H (1981) Cloned mouse mammary tumor virus DNA is biologically active in transfected mouse cells and its expression is stimulated by glucocorticoid hormones. Cell 23:335–345

Callahan R, Drohan W, Gallahan D, D'Hoostelaere L, Potter M (1982) Novel class of mouse mammary tumor virus-related DNA sequences found in all species of *Mus*, including mice lacking the virus proviral genome. Proc Natl Acad Sci USA 79:4113–4117

Cardiff RD, Wellings SR, Faulkin LJ (1977) Biology of breast preneoplasia. Cancer 39:2734–2746

Cardiff RD, Fanning TG, Morris DW, Ashley RL, Faulkin LJ (1981) Restriction endonuclease studies of hyperplastic outgrowth lines from BALB/cfC3H mouse hyperplastic mammary nodules. Cancer Res 41:3024–3029

Cohen JC (1980) Methylation of milk-borne and genetically transmitted mouse mammary tumor virus proviral DNA. Cell 19:653–662

Cohen JC, Murphey-Corb M (1983) Targeted integration of baboon endogenous virus in the BEVI locus on human chromosome 6. Nature 301:129–132

Cohen JC, Varmus HE (1979) Endogenous mammary tumour virus DNA varies among wild mice and segregates during inbreeding. Nature 278:418–422

Cohen JC, Varmus HE (1980) Proviruses of MMTV in normal and neoplastic tissues from GR and C3H mouse strains. J Virol 35:298–305

Cohen JC, Majors JE, Varmus HE (1979a) Organization of mouse mammary tumor virus-specific DNA endogenous to BALB/c mice. J Virol 32:483–496

Cohen JC, Shank PR, Morris VL, Cardiff R, Varmus HE (1979b) Integration of the DNA of mouse mammary tumor virus in virus-infected normal and neoplastic tissue of the mouse. Cell 19:333–345

Cohen JC, Breznik T, Gehrke CW, Ehrlich M (1981) In: Fields BN, Jaenisch R, Fox CF (eds) Animal virus genetics, ICN-UCLA symposia. Academic Press Differential methylation of endogenous and acquired mouse mammary tumor virus DNA. Mol Cell Biol 18:401–410

Cohen JC, Traina VL, Breznik T, Gardner M (1982) Development of a mouse mammary tumor virus-negative mouse strain: a new system for the study of mammary carcinogenesis. J Virol 44:882–885

Das MR, Mink MM (1979) Sequence homology of nucleic acids from human breast cancer cells and complementary DNAs from murine mammary tumor virus and Mason-Pfizer monkey virus. Cancer Res 39:5106–5113

Desrosiers RC, Mulder C, Fleckenstein B (1979) Methylation of Herpesvirus saimiri DNA in lymphoid tumor cell lines. Proc Natl Acad Sci USA 76:3839–3843

Dickson C, Smith R, Peters G (1981) In vitro synthesis of polypeptides encoded by the long terminal repeat region of mouse mammary tumour virus DNA. Nature 291:511–513

Donehower LA. Huang A, Hager GL (1981) Regulatory and coding potential of the mouse mammary tumor virus long terminal redundancy. J Virol 37:226–238

Drohan W, Schlom J (1979a) Diversity of mammary tumor viral genes within the genus Mus, the species Mus musculus and the strain C3H. J Virol 31:53–62

Drohan W, Schlom J (1979b) Differential distribution of mouse mammary tumor virus-related sequences in the DNAs of rats. JNCI 62:1279–1287

Drohan W, Teramoto YA, Medina D, Scholm J (1981) Isolation and characterization of a new mouse mammary tumor virus from BALB/c mice. Virology 114:175–186

Ehrlich M, Wang RY-H (1981) 5-Methylcytosine in eukaryotic DNA. Science 212:350–357

Etkind PR, Sarkar NH (1983) Integration of new endogenous mouse mammary tumor virus proviral DNA at common sites in the DNA of mammary tumors of C3Hf mice and hypomethylation of the endogenous mouse mammary tumor virus provirus DNA in C3Hf mammary tumors and spleens. J Virol 45:114–123

Etkind PR, Szabo P, Sarkar NH (1982) Restriction endonuclease mapping of the proviral DNA of the exogenous RIII murine mammary tumor virus. J Virol 41:885–867

Fanning TG, Puma JP, Cardiff RD (1980a) Identification and partial characterization of an endogenous form of mouse mammary tumor virus that is transcribed into the virion-associated RNA genome. Nucleic Acids Res 8:5715–5723

Fanning TG, Puma JP, Cardiff RD (1980b) Selective amplification of mouse mammary tumor virus in mammary tumors of GR mice. J Virol 36:109–114

Fanning TG, Vassos AB, Cardiff RD (1982) Methylation and amplification of mouse mammary tumor virus DNA in normal, premalignant, and malignant cells of GR/A mice. J Virol 41:1007–1013

Friedrich R, Morris VL, Goodman HM, Bishop JM, Varmus HE (1976) Differences between genomes of two strains of mouse mammary tumor virus as shown by partial RNA sequence analysis. Virology 72:330–340

Gillespie D, Gillespie S, Gallo RC, East JL, Dmochowski L (1973) Genetic origin of RD114 and other RNA tumour viruses assayed by molecular hybridization. Nature 244:51–54

Groner B, Hynes NE (1980) Number and location of mouse mammary tumor virus proviral DNA of normal tissue and of mammary tumors. J Virol 33:1013–1025

Groner B, Buetti E, Diggelmann H, Hynes N (1980) Characterization of endogenous and exogenous mouse mammary tumor virus proviral DNA with site-specific molecular clones. J Virol 36:734–745

Hayward WS, Neel BG, Astrin SM (1981) Activation of a cellular oncogene by promoter insertion in ALV-induced lymphoid leukosis. Nature 290:475–480

Herrlich P, Hynes NE, Ponta H, Rahmsdorf U, Kennedy N, Groner B (1981) The endogenous proviral mouse mammary tumor virus genes of the GR mouse are not identical and only one corresponds to the exogenous virus. Nucleic Acids Res 9:4981–4995

Heston WE (1958) Mammary tumors in agent-free mice. Ann NY Acad Sci 71:931–942

Heston WE, Deringer MK (1952) Test for a maternal influence in the development of mammary gland tumors in agent-free strain C3Hb mice. JNCI 13:167–175

Hilgers J, Bentvelzen P (1978) Interaction between viral and genetic factors in murine mammary cancer. Adv Cancer Res 26:143–189

Holliday R, Pugh JE (1975) DNA modification mechanisms and gene activity during development. Science 187:226–232

Hynes N, Groner B, Diggelmann H, Van Nie R, Michalides R (1979) Genomic location of mouse mammary tumor virus proviral DNA in normal mouse tissue and in mammary tumors. Cold Spring Harbor Symp Quant Biol 44:1161–1168

Hynes N, Kennedy N, Rahmsdorf U, Groner B (1981a) Hormone-responsive expression of an endogenous proviral gene of mouse mammary tumor virus after molecular cloning and gene transfer into cultured cells. Proc Natl Acad Sci USA 78:2038–2042

Hynes N, Rahmsdorf U, Kennedy N, Fabiani L, Michalides R, Nusse R, Groner B (1981b) Structure, stability, methylation, expression and glucocorticoid induction of endogenous and transfected proviral genes of mouse mammary tumor virus in mouse fibroblasts. Gene 15:307–317

Kennedy N, Kenediltschek G, Groner B, Hynes NE, Herrlich P, Michalides R, Van Ooyen AJJ (1982) Long terminal repeats of mouse mammary tumour virus contain a long open reading frame which extends into adjacent sequences. Nature 295:622–624

Kung HJ, Fung YK, Majors JE, Bishop JM, Varmus HE (1981) Synthesis of plus strands of retroviral DNA in cells infected with avian sarcoma virus and mouse mammary tumor virus. J Virol 37:127–138

Lasfargues EY, Lasfargues JC, Dion AS, Greene AE, Moore D (1976) Experimental infection of a cat kidney cell line with the mouse mammary tumor virus. Cancer Res 36:67–72

Lemons R, O'Brien SJ, Sherr CS (1977) A new genetic locus, BEVI, on human chromosome 6 which controls the replication of baboon type C virus in human cells. Cell 12:251–262

Lemons R, Nash WG, O'Brien SJ, Benveniste RE, Sherr CS (1978) A gene (BEVI) on human chromosome 6 is an integration site for baboon type C DNA provirus in human cells. Cell 14:995–1005

Long CA, Dumaswala UJ, Tancin SL, Vaidya AB (1980) Organization and expression of endogenous murine mammary tumor virus genes in mice congenic at the H-2 complex. Virology 103:167–177

Majors JE, Varmus HE (1981) Nucleotide sequences at host-proviral junctions for mouse mammary tumour virus. Nature 289:253–258

Mandel JL, Chambon P (1979) DNA methylation: organ-specific variation in the methylation pattern within and around ovalbumin and other chicken genes. Nucleic Acids Res 7:2081

McGhee JD, Ginder GD (1979) Specific DNA methylation sites in the vicinity of the chicken B-globin genes. Nature 280:419–420

McGrath CM, Marineau EJ, Voyles BA (1978) Changes in MuMTV DNA and RNA levels in BALB/c mammary epithelial cells during malignant transformation by exogenous MuMTV and by hormones. Virology 87:339–353

Medina D (1973) In: Busch H (ed) Preneoplastic lesions in mouse mammary tumorigenesis. Methods in cancer research. Academic, New York, pp 1–53

Michalides R, Schlom J (1975) Relationship in nucleic acid sequences between mouse mammary tumor virus variants. Proc Natl Acad Sci USA 72:4635–4649

Michalides R, van Deemter L, Musse R, van Nie (1978) Identification of the *Mtv-2* gene responsible for the early appearance of mammary tumors in the GR mouse by nucleic acid hybridization. Proc Natl Acad Sci USA 75:2368–2372

Michalides R, van Nie R, Nusse R (1981a) Mammary tumor induction loci in GR and DBAf mice contain one provirus of the mouse mammary tumor virus. Cell 23:165–173

Michalides R, Wagenaar E, Groner B, Hynes N (1981b) Mammary tumor virus proviral DNA in normal murine tissue and non-virally induced mammary tumors. J Virol 39:367–376

Morris VL, Medeiros E, Ringold GM, Bishop JM, Varmus HE (1977) Comparison of mouse mammary tumor virus-specific DNA in inbred, wild and Asian mice, and in tumors and normal organs from inbred mice. J Mol Biol 114:73–91

Morris VL, Kozak C, Cohen JC, Shank PR, Jolicoeur P, Ruddle F, Varmus HE (1979) Endogenous mouse mammary tumor virus DNA is distributed among multiple mouse chromosomes. Virology 92:46–55

Morse HC (ed) (1978) Origins of inbred mice. Academic, New York, pp 1–719

Nandi S, McGrath CM (1973) Mammary neoplasia in mice. Adv Cancer Res 17:353–414

Nusse R, Varmus HE (1982) Many tumors induced by the mouse mammary tumor virus contain a provirus integrated in the same region of the host genome. Cell 31:99–109

Nusse R, deMoes J, Hilkens J, van Nie R (1980) Localization of a gene for expression of mouse mammary tumor virus antigens in the GR/Mtv-2(−) mouse strain. J Exp Med 152:712–719

O'Brien JO (ed) (1980) Genetic maps, vol 1.

Parks WP, Scolnick EM (1973) Murine mammary tumor cell clones with varying degrees of virus expression. Virology 55:163–173

Payne GS, Bishop JM, Varmus HE (1982) Multiple arrangements of viral DNA and an activated host oncogene in bursal lymphomas. Nature 295:209–214

Peters G, Smith R, Brookes S, Dickson C (1982) Conservation of protein coding potential in the long terminal repeat of exogenous and endogenous mouse mammary tumor viruses. J Virol 42:880–888

Pitelka DR, DeOme KB, Bern HA (1960) Viruslike particles in precancerous hyperplastic mammary tissue of C3H and C3Hf mice. JNCI 25:753–777

Razin A, Riggs AG (1980) DNA methylation and gene function. Science 210:604–609

Riggs AD (1975) X-inactivation, differentiation and DNA methylation. Cytogenet Cell Genet 14:9–25

Ringold GM, Blair PB, Bishop JM, Varmus HE (1976) Nucleotide sequence homologies among mouse mammary tumor viruses. Virology 70:550–553

Ringold GM, Cardiff RD, Varmus HE, Yamamoto KR (1977a) Infection of cultured rat hepatoma cells by mouse mammary tumor virus. Cell 10:11–18

Ringold GM, Yamamoto KR, Shank PR, Varmus HE (1977b) Mouse mammary tumor virus DNA in integrated rat cells: characterization of unintegrated forms. Cell 10:19–26

Ringold GM, Shank PR, Yamamoto KR (1978) Production of unintegrated mouse mammary tumor DNA in infected rat hepatoma cells is a secondary action of dexamethasone. J Virol 26:93–101

Ringold GM, Shank PR, Varmus HE, Ring J, Yamamoto KR (1979) Integration and transcription of mouse mammary tumor virus DNA in rat hepatoma cells. Proc Natl Acad Sci USA 76:665–669

Shank PR, Cohen JC, Varmus HE, Yamamoto KR, Ringold GM (1978) Mapping of linear and circular forms of mouse mammary tumor virus DNA with restriction endonucleases: evidence for a large specific deletion occurring at high frequency during circularization. Proc Natl Acad Sci USA 75:2112–2116

Southern EM (1975) Detection of specific sequences among DNA fragments separated by gel electrophoresis. J Mol Biol 98:502–517

Sutter D, Doerfler W (1980) Methylation of integrated adenovirus type 12 DNA sequences in transformed cells is inversely correlated with viral gene expression. Proc Natl Acad Sci USA 77:253–256

Traina VL, Taylor BA, Cohen JC (1981) Genetic mapping of endogenous mouse mammary tumor viruses: locus characterization, segregation, and chromosomal distribution. J Virol 40:735–744

Van Nie R, de Moes J (1977) Development of a congeneic line of the GR mouse strain without early mammary tumours. Int J Cancer 20:588–594

Van Nie R, Verstraeten AA (1975) Studies of genetic transmission of mammary tumour virus by C3Hf mice. Int J Cancer 16:922–931

Van Nie R, Verstraeten AA, deMoes J (1977) Genetic transmission of mammary tumour virus by GR mice. Int J Cancer 19:383–390

Varmus HE, Bishop JM, Nowinski RC, Sarkar, NH (1972) Mammary tumour virus-specific nucleotide sequences in mouse DNA. Nature 238:189–190

Varmus HE, Quintrell N, Ortiz S (1981) Retroviruses as mutagens: insertion and excision of a nontransforming provirus alter expression of a resident transforming provirus. Cell 25:23–36

Verstraeten AA, van Nie R (1978) Genetic transmission of mammary tumour virus in the DBAf mouse strain. Int J Cancer 21:473–475

Weiss R, Teich N, Varmus H, Coffin J (eds) (1982) RNA tumor viruses. Cold Spring Harbor Press, New York

# Mouse Mammary Tumor Virus Expression and Mammary Tumor Development

Rob Michalides, Albert van Ooyen, and Roeland Nusse

## 1 Introduction

Mammary tumors in mice can be induced by a milk-transmitted virus, the mouse mammary tumor virus (MMTV). Since this observation (Bittner 1936), many aspects of virus-induced mammary tumorigenesis have been studied, the idea being that this system may serve as a general model for mammary tumorigenesis. Numerous studies have described the route of infection, the structure of the virus, and virus prevalence among inbred strains of mice (for review see Bentvelzen and Hilgers 1980). Until recently, however, no clues were provided as to how MMTV could transform mammary gland cells into mammary tumors. Recent insights into the life cycle of retroviruses and the detailed analysis of the structure of the MMTV genome are now leading to an initial understanding of the mechanism of MMTV-induced mammary tumorigenesis.

MMTV belongs to the group of slowly transforming retroviruses. It induces mammary tumors in susceptible mice after a latency period of about 1 year. Factors such as virulence of the MMTV variants, histocompatibility genes of the host, and the hormonal status of the animal influence the onset of mammary tumors and mammary tumor incidence (for review see Hilgers and Sluyser 1981).

Department of Virology, Antoni van Leeuwenhoekhuis, The Netherlands Cancer Institute, Plesmanlaan 121, 1066CX Amsterdam, The Netherlands

Current Topics in Microbiology and Immunology, Vol. 106
© Springer-Verlag Berlin·Heidelberg 1983

The mammary gland is the main target of MMTV. Whereas most of the mammary gland cells do become infected with MMTV, only one or a few mammary tumors appear in the MMTV-infected animal. Mammary glands contain alveolar, ductal, and myoepithelial cells in a branching system of ducts that terminate in clusters of alveoli, which secrete milk during lactation. Myoepithelial cells, which contract during suckling, form a layer around the ducts and alveoli. In one mouse strain, C3H, only alveolar cells become transformed by MMTV (SMITH et al. 1980), whereas in the GR strain, both ductal and alveolar cells are transformed (VAN NIE 1981).

MMTV can be transmitted in two ways, the first being via the milk from mother to offspring. The MMTV transmitted in this way is called exogenous MMTV. Retroviruses replicate via a provirus DNA intermediate, which becomes integrated in the cellular DNA of the host cell. Therefore, only cells infected with exogenous MMTV contain proviral DNA of exogenous MMTV. The second method of transmission is genetically via the gametes. Every inbred mouse strain tested so far has some MMTV proviral DNA copies integrated in the germ line. These so-called endogenous MMTVs are transmitted to the offspring via the gametes and are therefore present in all mouse cells. Both exogenous and endogenous MMTVs are implicated in the development of mammary tumors. The mechanism of transformation by these two differently transmitted MMTVs may be identical, in that nonselective integrations of DNA from either exogenous MMTV or activated endogenous MMTV into the host cell DNA may affect cellular genes crucial to the development of mammary tumors.

# 2 Structure and Expression of the Mouse Mammary Tumor Virus Genome

## 2.1 Structure of the Genome

MMTV contains an RNA with a sedimentation coefficient of 60–70 $S$ and a molecular mass of $6.45 \times 10^6$ daltons (DUESBERG and CARDIFF 1968; DION et al. 1977). The native MMTV RNA consists of two subunits of 35 $S$, each with a molecular mass of $2.93 \times 10^6$ daltons. These subunits may be held together at the 5′ ends to form a 60–70 $S$ dimer, as is described for other retroviruses (BISHOP 1978). The 3′ termini of the MMTV RNA are polyadenylated (SCHLOM et al. 1973a), and the 5′ termini likely carry the CAP structure, 5′ $m^7GpppG$-, which is present in the RNAs of other retroviruses (BISHOP 1978). Like all retroviruses, MMTV RNA replicates via a DNA intermediate, using an RNA-dependent DNA polymerase, and becomes integrated in the cellular DNA in a MMTV provirus form. Unintegrated MMTV DNA is persistently present in nonmurine cells infected with MMTV; in murine cells infected with MMTV, only the integrated form of MMTV DNA can be found (VAIDYA et al. 1976; RINGOLD et al. 1977). Unintegrated MMTV DNA in MMTV-infected rat hepatoma cells is present as linear or open circular duplexes containing two LTRs (long terminal repeats), and also as covalently closed circular viral DNA, with either one or two LTRs (RINGOLD et al. 1977; SHANK et al. 1978). The unintegrated MMTV DNAs from rat hepatoma cells infected with MMTV(GR) (SHANK et al. 1978) and from mink lung cells infected with MMTV(C3H) (COHEN et al. 1979b) have been isolated and restriction enzyme maps of them have been made. The MMTV DNAs of these two different MMTVs appeared highly related; only two out of twelve restriction endonucleases tested recognized different sites within the MMTV DNAs. The high degree of homology between the RNA genomes of MMTV(GR) and MMTV(C3H) was also demonstrated in earlier experiments using competition hybridization tests (MICHALIDES and SCHLOM 1975).

Fragments of these unintegrated MMTV DNAs have been cloned (GRONER et al. 1980; MAJORS and VARMUS 1981) and used for further studies. In only one case, a whole genome-length MMTV DNA containing one LTR was cloned in a permutated form (BUETTI and Diggelman 1981). This MMTV(GR)DNA was cloned at a unique EcoRI site, by which a sequence permutation with respect to linear MMTV DNA is produced. This permutated MMTV DNA was ligated to itself in order to obtain MMTV sequences in the right order and was then transfected to LTK⁻ cells, together with TK DNA. The TK⁺ cell clones transfected with reconstructed MMTV DNA synthesized normal viral RNA and proteins; the viral gene expression was increased by the addition of dexamethasone (BUETTI and DIGGELMAN 1981; OWEN and DIGGELMAN 1983). This finding is the more striking, since many attempts to clone a whole integrated MMTV provirus have been unsuccessful (MAJORS and VARMUS 1981; DONEHOWER et al. 1981), with two exceptions (HYNES et al. 1981; DIGGELMAN et al. 1982). Numerous attempts to clone a complete integrated copy of exogenous or endogenous MMTVs have failed. When EcoRI-digested cellular DNA was used to isolate right and left halves of MMTV proviruses, usually only clones with right halves and their flanking cellular DNA were identified. This has led to the assumption that the left half of the MMTV genome contains a sequence that cannot readily be cloned in Escherichia coli vector systems. Such a poisonous sequence would be absent from the unintegrated MMTV genome cloned by BUETTI and DIGGELMAN (1981), and also from the clones containing an endogenous MMTV genome from GR DNA (HYNES et al. 1981) and from A/J DNA (DIGGELMAN et al. 1982). The MMTV proviruses of these last two clones may well be identical, since they show restriction enzyme maps identical to MMTV provirus Mtv-8 (see Sect. 3). This Mtv-8 may lack the poisonous sequence. Upon transfection, the cloned endogenous MMTV provirus of Mtv-8 produces the correct 35 S and 24 S MMTV RNAs. The viral gene expression is stimulated by the addition of dexamethasone, and precursor MMTV proteins are being synthesized. However, no virus particles are being produced, which may suggest that the Mtv-8 MMTV provirus is partially defective.

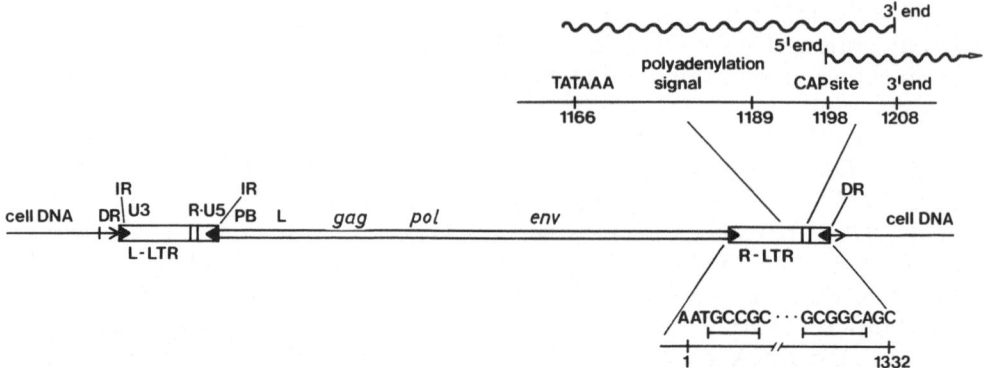

**Fig. 1.** Structure of an integrated MMTV provirus. The numbers at the R-LTR indicate the nucleotide positions of signal sequences from the left of the R-LTR. *DR*, direct cellular repeat of six nucleotides (NT); *IR*, inverted repeat (6 NT); *U3*, unique 3′ region LTR; *R*, repeat region; *U5*, unique 5′ region LTR; *PB*, binding site for tRNA primer; *L*, leader region; *gag, pol, env*, structural genes

Using cloned subfragments of MMTV DNAs and of integrated MMTV proviruses, extensive nucleotide sequencing studies have revealed control signals on the MMTV provirus and the coding potential of the MMTV genome (MAJORS and VARMUS 1981; KLEMENZ et al. 1981; DONEHOWER et al. 1981; FASEL et al. 1982; KENNEDY et al. 1982), which is summarized in Fig. 1. The integrated MMTV provirus shows the following features:

*Direct Cellular Repeat Region (DR).* The integration site for exogenous MMTV DNA is not specific (see Sect. 5). Numerous sites in the cellular DNA can therefore be used for the integration of MMTV DNA. At both sides of an integrated exogenous MMTV provirus, and also at the endogenous MMTV provirus *Mtv-8*, one finds a direkt cellular repeat of six nucleotides which is different for each integration site. This direct repeat sequence is present only once in the integration site before insertion; the duplication is therefore generated during the integration process, likely by an endonuclease activity which produces staggered ends at the cellular DNA (MAJORS and VARMUS 1981). DRs in host DNA sequences have also been found next to several transposable elements, suggesting a similarity between retroviruses and transposable elements.

*Long Terminal Repeat (LTR).* LTRs are found at each end of the integrated MMTV provirus. They are generated at the transcription of RNA into DNA (GILBOA et al. 1979). Specific fragments of the MMTV RNA genome are duplicated within the LTR at the opposite side of the provirus. An unique region of MMTV RNA at the 3′ end, the U3, becomes duplicated in the left LTR, and a unique region at the 5′ end, U5, duplicates in the right LTR. Each LTR unit therefore contains the elements IR, U3, R,

U5, and IR. MMTV DNA contains an extraordinarily long LTR sequence of 1332 nucleotides.

*Inverted Repeat* (IR). The left and right ends of each LTR unit contain an inverted repeat of six bases (AA*TGCCGC...GCGGCA*GC), starting two base pairs from the ends. The two base pairs at the left end of the left LTR and at the right of the right LTR are lost during integration of the MMTV provirus.

*Unique 3' Region (U3).* This region, present at the 3' end of the MMTV-RNA genome, is defined as the region between the primer binding site for plus-strand DNA synthesis and the short repeat region near the 3' end of the genome. The U3 region of MMTV LTR, with a length of approximately 1190 nucleotides, contains several regulatory signals and also an open reading frame.

*Short Repeat Region (R).* At both ends of the viral genome there is a short repeat region. From studies determining the 3' and 5' ends of the viral RNAs, it is concluded that R contains approximately ten nucleotides (KLEMENZ et al. 1981; VAN OOYEN et al. 1983).

*Unique 5' Region (U5).* The U5 region is present at the 5' end of the viral genome between R and the primer binding site for minus-strand DNA synthesis.

*Primer Binding Site for Minus-Strand DNA Synthesis (PB).* The proviral sequence to the right of the left LTR contains a 16-base-pair sequence matching the 3' end of tRNA$^{lys}$ which functions as a primer for the minus-strand DNA synthesis of MMTV DNA (PETERS and GLOVER 1980; WATERS 1981).

*Leader Region (L).* The leader region is an untranslated sequence preceding the coding region for structural proteins.

*Gag, pol, and env.* Structural genes code for the internal proteins (*gag*), RNA-dependent DNA polymerase (*pol*), and envelope (*env*) of MMTV.

The LTR unit contains several regulatory signals, such as a Goldberg-Hogness TATAAA sequence at position 1166, a binding site for the RNA polymerase, and a polyadenylation signal, AGTAAA, at position 1189. The 5' CAP site of MMTV RNA is probably located around position 1198 and the 3' end of the viral RNA genome around position 1208 (see Fig. 1). Since the polyadenylation signal is located between the TATAAA box and the CAP site, this signal is not transcribed in a starting MMTV RNA molecule. The starting MMTV RNA may therefore bypass the 3' end around position 1208 without being terminated. The U3 region also contains an interaction site where the dexamethasone-receptor couples, as discussed elsewhere in this volume by RINGOLD et al. Nucleotide sequencing of the LTR of MMTV (DONEHOWER et al. 1981; KENNEDY et al. 1982; FASEL et al. 1982) has revealed a conserved open reading frame in the right LTR of 960 nucleo-tides, with a coding potential for a basic 36 700 dalton protein. Its structure, expression, and possible function are discussed in the following section.


## 2.2  Expression of the Genome

MMTV-producing cells derived from murine mammary tumors or MMTV-infected nonmurine cells have been used to study the expression of MMTV (for review see MICHALIDES and NUSSE 1981; WEISS et al. 1982). Two major classes of RNA are detected in MMTV-producing cells: a full-length ge-nome-sized 35-$S$ RNA of 7.8 kb and a 24-$S$ RNA of 3.8 kb (ROBERTSON and VARMUS 1981; DUDLEY and VARMUS 1981) (see Fig. 2). Synthesis of these RNAs in vitro is stimulated by the addition of dexamethasone. After dexamethasone stimulation, the first augmented RNA is the 35-$S$ MMTV RNA, which is probably the precursor to the 24-$S$ MMTV RNA. The 24-$S$ MMTV RNA is then derived from the 35-$S$ RNA by a splicing mecha-nism, at which time the *gag-pol* information is removed from the 35-$S$ MMTV RNA. The 5′ splice site is located 153 nucleotides to the right of the left LTR, as determined by S1 analysis (VAN OOYEN et al. 1983) (see Fig. 2). Various studies described elsewhere in this volume and in *Micha-lides* and *Nusse* (1981), demonstrated that the 35-$S$ MMTV RNA instructs the synthesis of a precursor protein of *gag* and *pol*, whereas the 24-$S$ MMTV RNA codes for the *env* protein. The *env* gene has now been localized on the MMTV genome by comparing the amino acid sequences of the *env* gene products with the nucleic acid sequence of cloned MMTV DNA (Ma-jors and Varmus, personal communication). The amino terminal end of gp52 is located approximately 0.8 kb toward the 3′ end of the internal *Eco*RI cleavage site (at approximately 5.2 kb from the 5′ end of the MMTV provirus); the first AUG of gp52 starts one nucleotide after the 3′ splice site of 24-$S$ RNA. The carboxyl terminal end of gp 36 is located in the LTR, and 18 amino acids at this end are coded for by the 5′ end of LTR. This leaves no coding sequences between *env* and LTR, as assumed by others (BENTVELZEN and HILGERS 1980). The *gag-pol* gene has not yet been mapped on the MMTV genome, since its amino and carboxyl termini are not known and its corresponding MMTV DNA regions have not yet been sequenced. The coding region for *gag-pol* in Fig. 2 is, therefore, a presumed one.

Recently, DNA sequencing has revealed an open reading frame (*orf*) in the LTR of MMTV (DONEHOWER et al. 1981; FASEL et al. 1982; KENNEDY et al. 1982). Its presence was confirmed by in vitro translation of 3′ frag-ments from MMTV poly(A)$^+$ RNA (DICKSON and PETERS 1981; DICKSON et al. 1981). The conservation of the LTR sequence in various MMTVs suggests a role for the orf protein and has prompted a search for *orf* mRNA. Northern blot analysis, combined with nuclease S1 mapping data, has re-vealed the presence of a poly(A)$^+$ RNA of about 1.7 kb, which is a candi-date mRNA for the LTR gene product (VAN OOYEN et al. 1983; WHEELER

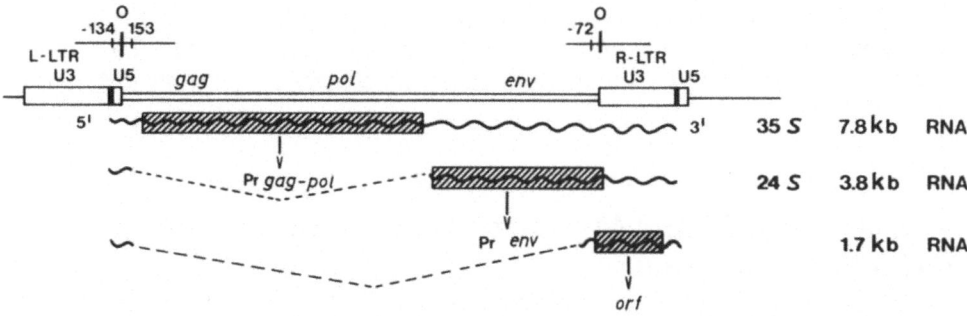

**Fig. 2.** Expression of the MMTV provirus. The numbers at the L-LTR indicate the 5′ start site of the MMTV RNAs (−134 nucleotides from the right of the L-LTR) and the 5′ splice site of the MMTV RNAs (153 nucleotides from the right of the L-LTR). Position −72 at the R-LTR indicates the 3′ splice site (−72 nucleotides from the left of the R-LTR) of the 1.7-kb RNA. ▨, protein-coding region; ---, spliced region; *orf,* open reading frame

et al. 1983). The RNA is prominent in Balb/c mammary glands and mammary tumors but can also be found in C3H and GR, albeit at a much lower level. Nuclease S1 analysis showed that the 5′ end of this RNA maps at a position indistinguishable from 35-*S* and 24-*S* MMTV RNA, about 130 nucleotides upstream from the 3′ end of the left LTR (see Fig. 2). Colinearity with proviral DNA continues to a position about 153 nucleotides downstream from the left LTR and is resumed 72 nucleotides upstream from the right LTR. Nucleotide sequence analysis revealed the presence of a splice acceptor consensus sequence at this position. No proviral sequences of the *gag, pol,* and *env* regions are present. Termination of transcription probably occurs at the normal site in the LTR (KLEMENZ et al. 1981). The composition of 1.7-kb RNA can be estimated as follows: (a) 1330 nucleotides of LTR sequence, (b) 10–14 nucleotides of repeat sequences, (c) 227 nucleotides of non-LTR MMTV-specific sequences, and (d) a polyA tail of 100–150 nucleotides (EDMONDS and CARAMELA 1969; LIM and CANELLAKIS 1970). This totals 1.7 kb, which is in agreement with RNA blotting data.

The anatomy of *orf* mRNA suggests that it may arise as a splice product. Another possibility is that *orf* mRNA is colinear with an incomplete endogenous provirus, such as found in Balb/c (COHEN et al. 1979a). The latter would explain the high level of orf mRNA in Balb/c mammary glands and tumors (VAN OOYEN et al. 1983; WHEELER et al. 1983) and is in agreement with the observation that the subgenomic copy of MMTV in Balb/c (unit I) is hypomethylated in Balb/c mammary tumors (BREZNIK and COHEN 1982). It is also consistent with the low level of *orf* mRNA in strains GR and C3H. MMTV unit I is absent in these strains.

The structure of the MMTV provirus is, in general, similar to that of other replication-competent retroviruses; it contains two LTRs and the structural genes, *gag, pol,* and *env*. The structure shows one unique feature: the conserved orf region in the U3 part of the LTR, which is also expressed as a separate mRNA. The expression of the MMTV provirus may be influenced at the various levels of transcription, translation, and the processing of the precursor proteins, and changes in the expression of MMTV proviruses have been linked with mammary tumorigenesis, as will be discussed in Sect. 4.


# 3 Endogenous Mouse Mammary Tumor Virus Proviruses and *Mtv* Induction Loci


MMTV can be transmitted by two routes: genetically, as an endogenous provirus, and by infection via a milk borne exogenous virus. The endogenous proviruses are present in the DNA of the germ line and of somatic cells of the mouse, segregating as an inherited trait. Endogenous MMTV proviruses are found in all mouse strains tested (VARMUS et al. 1972; MICHALIDES and SCHLOM 1975) but are absent, or only heterozygously present, in feral mice (COHEN and VARMUS 1979; CALLAHAN et al. 1982). The inbred strains of mice examined also vary with respect to the number of MMTV proviral genomes in their germ line DNA and their location in the cellular genome. These findings led to the idea that genetically transmitted MMTV proviruses are the result of infrequent insertions of infectious MMTV genome into the germ line during evolution of the species (COHEN and VARMUS 1979). CALLAHAN et al. (1982) found, by relaxing the hybridization conditions, that MMTV-negative mice and feral mice contain MMTV-related sequences, called MMTV-$\beta$ sequences. These MMTV-$\beta$ sequences are ubiquitously present in the genus *Mus* and lack variation in the pattern of restriction fragments. MMTV-$\beta$ may therefore well be regarded as the evolutionary progenitor of the infectious MMTV present in inbred strains of mice, which is now called MMTV-$\alpha$. Whether MMTV-$\beta$ is involved in the development of mammary tumors is yet unknown. The exogenously transmitted MMTV of MMTV-$\alpha$ is, in general, a causative agent in mammary tumor development, and, in two cases, endogenous proviruses of MMTV-$\alpha$ have been implicated in mammary tumorigenesis (MICHALIDES et al. 1981a). The endogenous MMTV proviruses of MMTV-$\alpha$ are classified by TRAINA et al. (1981), using the characteristic *Eco*RI DNA fragments of mouse DNA containing MMTV DNA sequences. *Eco*RI cleaves at only one site within most of the nondefective viral genomes. Cellular DNA fragments thereby contain one half of the MMTV proviral DNA and a cellular DNA sequence up to the first *Eco*RI recognition site in the flanking cellular DNA. This yields well-defined *Eco*RI fragments of cellular DNA fragments containing MMTV DNA. By analysis of recombinant inbred strains, TRAINA et al. (1981) described 15 loci containing MMTV proviral DNA information, called *Mtv* loci. Two of these loci, *Mtv-1* and *Mtv-2,* are directly related to mammary tumor development (MICHALIDES et al. 1981a).

In case of the *Mtv-1* and *Mtv-2* loci, expression of the MMTV provirus is necessary for mammary tumor development. Two other *Mtv-* loci, *Mtv-3* and *Mtv-5*, are expressed, but this expression is not linked with mammary tumorigenesis. Expression of the other *Mtv* loci has not yet been studied.

# 4  Mouse Mammary Tumor Virus Expression in Mammary Glands and Mammary Tumors

## 4.1  Expression in Nonvirally Induced Mammary Tumors

Since biological studies clearly indicate a causative role for exogenous MMTV in the development of mammary tumors, and expression of particular endogenous MMTVs in mouse strains GR and C3Hf is directly linked with mammary tumorigenesis (MICHALIDES et al. 1981a), it is an obvious assumption that expression of MMTV may be linked with mammary tumorigenesis in general, and that control over expression of MMTV proviruses is essential in this process. Therefore many comparative studies have been performed, describing levels of MMTV RNA and proteins in a variety of systems, searching for direct correlations between MMTV expression and mammary tumor development. Generally, mouse strains with a high incidence of mammary tumors, due to either milk-borne or genetically transmitted MMTV, express high levels of MMTV RNA and proteins in their mammary glands and mammary tumors (AXEL et al. 1972; SCHLOM et al. 1973b; VARMUS et al. 1973; McGRATH et al. 1978; MICHALIDES et al. 1978). Up to approximately 0.5% of the total RNA from mammary tumors may be MMTV RNA, corresponding to approximately 1000 virus RNA copies per cell. Nontarget tissues, such as spleen and liver, contain very low levels (less than one copy per cell) (VARMUS et al. 1973). Mammary glands of virus-free animals contain low levels of MMTV RNA (MICHALIDES and NUSSE 1981). Mammary tumors can be induced in these low-mammary-tumor strains by hormones, carcinogens, X-irradiation, or combinations of these treatments. This led to many investigations to look for MMTV expression in these nonvirally induced mammary tumors. The mouse strain Balb/c was, in particular, used in these studies. This mouse strain has a low mammary tumor incidence (approximately 30%) in breeding females older than 1 year. Prolonged hormonal stimulation of mammary glands, achieved by implanting pituitary isografts in the kidneys, increases the incidence of mammary tumors. Mammary tumors can also be induced by urethane, dimethylbenzanthracene, and X-irradiation, or by combination of the hormones, chemicals, and irradiation (MICHALIDES et al. 1979; BUTEL et al. 1981). These hormonal and chemical treatments induce preneoplastic lesions in mammary glands, termed hyperplastic alveolar nodules (HAN). The HANs can be isolated and reimplanted in virgin mice, where they form mammary tumors (MEDINA 1978). They are therefore regarded as intermediates between normal mammary glands and mammary tumors.

Normal Balb/c mammary glands hardly contain any MMTV RNA (MI-

CHALIDES et al. 1978; DUDLEY et al. 1978b). Hormones, which affect the development of the mammary glands, do have a slight effect on the levels of MMTV RNA; higher levels of MMTV RNA and proteins were found in lactating mammary glands of young Balb/c females (McGRATH et al. 1978) and in mammary glands of animals treated with prolactin (MICHA-LIDES et al. 1978, 1979). Some outgrowths of HANs, induced either by hormones or by chemicals, do contain detectable levels of MMTV RNA, ranging from three to nine MMTV RNA copies per cell, whereas other outgrowths do not contain any MMTV RNA (DUDLEY et al. 1978b; DUSING-SWARTZ et al. 1979). Cell lines from these outgrowths show the same picture; some express MMTV RNA, whereas others lack MMTV expression (DUD-LEY and BUTEL 1979; PAULEY et al. 1979). The expression in MMTV-positive HANs or cell lines derived from them is stimulated two-to fourfold by dexamethasone treatment, suggesting that transcription can be stimulated by glucocorticoid hormones. Hormonally and chemically induced Balb/c mammary tumors contain enhanced levels of MMTV RNA (MICHALIDES et al. 1979; McGRATH et al. 1978; BUTEL et al. 1981). Analysis of individual chemically induced tumors for the presence of MMTV RNA showed also MMTV-RNA-negatice mammary tumors (BUTEL et al. 1981; DUSING-SWARTZ et al. 1979).

MMTV expression in induced mammary tumors and HANs in Balb/c appears incomplete. A preferential transcription of polyA-adjacent sequences of MMTV RNA was observed (DUDLEY et al. 1978a; BUTEL et al. 1981). These tissues also contain only reasonable amounts of the 28000-dalton major MMTV core protein (p 28), but no immunologically detectable 52000-dalton MMTV envelope glycoprotein (gp 52) (TERAMOTO et al. 1980; McGRATH et al. 1981), suggesting a noncoordinate expression of MMTV gene products. Balb/c mammary gland tissues contain 35-$S$ MMTV RNA, coding for the *gag-pol* protein; 24-$S$ MMTV RNA, coding for the envelope proteins; and a novel 1.7-kb RNA, which contains sequences from the LTR (WHEELER et al. 1983; VAN OOYEN et al. 1983). Preneoplastic tissues and mammary tumors of Balb/c contain less 35-$S$ and 24-$S$ MMTV RNA, but still relatively high levels of 1.7-kb RNA. The absence of envelope proteins in Balb/c mammary gland tissues may therefore be due to a defect in translation. The persistent expression of the 1.7-kb RNA in Balb/c mammary tumors may well represent the expression of the incomplete endogenous MMTV provirus in Balb/c DNA; this is also suggested by its hypomethylation in Balb/c mammary tumors (BREZNIK and COHEN 1982).

The preferential transcripts of polyA-adjacent sequences of MMTV RNA in Balb/c mammary tumors observed by DUDLEY et al. (1978a) correspond, most likely, with the 1.7-kb LTR-specific RNA found by VAN OOYEN et al. (1983) and WHEELER et al. (1983).

The conclusion from these studies is that treatments of mice with hormones and carcinogens result in the development of mammary tumors, some of which express moderate amounts of MMTV RNA, whereas only very low amounts of MMTV RNA (less than one MMTV RNA copy per cell) are found in normal mammary glands of Balb/c mice. Hormones and carcinogens, therefore, do have an effect on the transcription of endogenous MMTV proviruses. The induced MMTV expression is, however, not directly correlated with maintenance, and probably also not with the initiation of mammary tumor development. This can be said because part of the induced HANs or mammary tumors do not contain any MMTV RNA at all, and because the induced mammary tumors usually contain less MMTV RNA than mammary glands of the hormonally or chemically treated animals.

Similar results have been obtained with other low-mammary-tumor strains, such as O20, C57BL, and C3Hf (NUSSE et al. 1980; MICHALIDES et al. 1978) and DD, KF, and DDY mouse strains (IMAI et al. 1982).

Another way of searching for a possible role of MMTV in the development of nonvirally induced mammary tumors is by looking for extra MMTV proviral DNA sequences in Southern DNA blots of these tumors. The presence of extra MMTV proviral sequences in the tumor DNA indicates an integration of MMTV DNA from an activated endogenous MMTV provirus. Only some of the hormone or  chemically induced mammary tumors contain extra MMTV DNA information, suggesting that reintegration of MMTV DNA in these tumors is not a prerequisite for tumor formation (MICHALIDES et al. 1981 b; DROHAN et al. 1982).

## 4.2  Proviral Methylation and Mouse Mammary Tumor Virus Expression

DNA from a broad spectrum of eukaryotes contains 5-methylcytosine, in addition to the normal bases (for references see BIRD and SOUTHERN 1978). Modification occurs almost exclusively at the sequence CpG. Studies on DNA methylation make use of restriction enzymes which contain this doublet in their recognition sequence. The combination *Hpa*II/*Msp*I has especially been useful. Both enzymes recognize the sequence CCGG. *Hpa*II is inhibited by methylation at the C residue, while *MSp*I is not. In general, these studies have revealed that transcriptionally active genes are less methylated than their inactive counterparts (MANDEL and CHAMBON 1979).

The 5-methylcytosine content of MMTV-specific DNA sequences acquired by both milk-borne infection and genetic transmission was determined by COHEN (1980). In the liver, which is resistant to milk-borne MMTV infection (COHEN et al. 1979b), extensive methylation of proviral DNA was found. Acquired MMTV proviruses, however, both in tumors and in normal lactating mammary glands, were specifically hypomethylated. Earlier studies on MMTV RNA synthesis (VARMUS et al. 1973; McGRATH et al. 1978) have revealed that the concentration of viral RNA in the liver of both uninfected and infected mice is very low. In infected tissues, either normal or transformed, a high level of MMTV RNA can be found. Thus there is a direct correlation between the demethylation of bases in MMTV-specific proviral sequences and their expression. Similar conclusions were drawn by FANNING et al. (1982). Subsequent studies have demonstrated that demethylation of endogenous proviral DNA may also occur. The subgenomic unit I in Balb/c mice may become demethylated during mammary carcinogenesis (BREZNIK and COHEN 1982). Also, some endogenous MMTV sequences are specifically hypomethylated in spontaneously occurring or chemically induced C3H/Sm mammary tumors (DROHAN et al. 1982). A complicating factor in these studies is that several copies of MMTV proviral DNA are present in all mouse strains investigated. However, transcription from individual proviruses was not measured separately. Thus, no direct links can be made between demethylation of certain proviruses and their expression.

## 4.3 Mammary Tumor Induction Loci *Mtv-1* and *Mtv-2*

The expression of MMTV has, in two cases, been linked with the particular endogenous MMTV proviruses, *Mtv-1* and *Mtv-2*, and was also found to correlate with mammary tumor development in the strains of mice involved. The *Mtv-1* locus is responsible for expression of moderate levels of MMTV and for late-appearing mammary tumors in the genetically related C3Hf and DBAf mouse strains (VAN NIE and VERSTRAETEN 1975). A particular endogenous MMTV provirus was associated with the *Mtv-1* locus by relating the presence of MMTV in the milk with segregating MMTV-specific DNA restriction fragments of individual mice of the backcross $Mtv-1 (-/-)$ × $Mtv-1 (-/+)$ (MICHALIDES et al. 1981a). C3Hf and DBAf mice have five endogenous MMTV proviruses, one of which, characterized by EcoRI fragments of 5.9 and 4.2 kb, is associated with the *Mtv-1* locus. Similarly, one of the five endogenous MMTV proviruses of GR mice was linked with the *Mtv-2* locus. This dominant locus controls expression of high levels of MMTV in the milk and the occurrence of pregnancy-dependent mammary tumors in GR mice (VAN NIE et al. 1977). Similar backcross studies, as performed with DBAf mice, identified the MMTV provirus, characterized by EcoRI fragments of 11.0 and 6.9 kb, to correspond with the *Mtv-2* locus. Two congeneic strains have been made, one identical to GR but lacking the *Mtv-2* locus, so-called GR-$Mtv-2^-$ (VAN NIE and DE MOES 1977) and one acquiring this locus, O20-$Mtv-2^+$ (VAN NIE in preparation). The lack of or acquisition of the *Mtv-2* locus in both congeneic strains did correspond with the respective loss of or gain of the 11.0 and 6.9 kb EcoRI MMTV DNA fragments (MICHALIDES et al. 1981a). These findings clearly indicate that expression of two endogenous MMTVs is linked with mammary tumor formation. The next question then becomes: Is expression of these endogenous MMTV proviruses, by itself, sufficient for mammary tumor formation, or is reintegration of MMTV DNA a necessary step? This has mainly been studied in the GR mouse strain. Almost all females of this strain develop pregnancy-dependent mammary tumors before 9 months of age. These mammary tumors have a characteristic morphology and are classified as plaque-type tumors (FOULDS 1975). GR mouse mammary tumors contain only a few extra MMTV DNA copies, compared with liver DNA. Most of them are present at submolar concentrations, suggesting that not all the tumor cells contained the extra MMTV proviral copies (COHEN and VARMUS 1980; MICHALIDES et al. 1981 b). Using restriction enzymes, which specifically detect the *Mtv-2*-associated MMTV provirus, FANNING et al. (1982) were able to demonstrate that mammary glands of GR mice contain additional copies of the *Mtv-2*-associated MMTV provirus. Premalignant outgrowths and tumor cells of GR mice contained an average of three additional *Mtv-2*-associated MMTV proviruses. Using different restriction enzymes, it was found that premalignant outgrowths and mammary tumors consisted largely of heterogeneous cell populations. This was also described by MACINNES et al. (1981) and MICHALIDES et al. (1982a), who used serially transplanted GR mouse mammary tumors. Primary mammary tumors in this strain are hormone dependent and convert, upon serial transplantation, to autonomous, hormone-independent growth. Extra MMTV DNA copies, which were present in the primary tumor and in the initially hormone-dependent tumor passages, disappeared from the finally hormone-independent passages of a particular tumor line (MICHALIDES et al. 1982a). This is explained by assuming that the original tumor contained both hormone-dependent and hormone-independent cells and that the hormone-dependent cells disappeared from the tumor at serial transplantation, due to reduced growth rate (SLUYSER et al. 1976).

It therefore appears that the *Mtv-2*-associated MMTV provirus becomes expressed in GR mouse mammary glands and becomes integrated at different sites of the cellular genome. Mammary tumors in GR mice consist of heterogeneous cell populations, each of which may contain different extra MMTV proviruses. The overall DNA from GR mouse mammary tumors, then, contains extra MMTV DNA copies at submolar quantities. This is in contrast with milk-borne-virus-induced mammary tumors in female Balb/cfC3H, which are of clonal origin (see also Sect. 5). Repositioning of MMTV provirus may therefore be required for the full expression of oncogenic potential of endogenous MMTV DNA. The GR mouse mammary tumor may be exceptional, in that it contains multiple initially transformed cells with integrations of MMTV DNA at different sites.

# 5 Acquisition of the Mouse Mammary Tumor Virus Genome During Tumorigenesis

In the previous sections we have outlined how the viral genome of MMTV is organized and how it is expressed into its gene products. This section will deal with the mechanism of oncogenesis by MMTV, with emphasis on the recent discovery that the integrated proviral genome can act as an insertional mutagen.

The viral genome of MMTV does not seem to be equipped with its own oncogene. This can be concluded from the inability of this virus to transform cells in vitro and also from the extensive mapping and sequencing of the genome itself. There is, however, one gene whose function still has to be defined: the large open reading frame on the LTR. This sequence is not derived from the host genome, as is the case with the *onc* genes of the rapidly transforming retroviruses, which makes it an unlikely candidate for a transforming gene.

MMTV is not alone in lacking a viral oncogene. Viruses such as the avian leukosis virus and the murine leukemia virus are also devoid of transforming sequences. Also the biological properties of these viruses have a resemblance to MMTV: tumors appear late after administration of the virus (6–12 months), and the tumors are clonal, with respect to the integrated viral genome. In contrast, tumors induced by transforming retroviruses develop within a couple of weeks and are derived from many cells, with the viral genome integrated at different sites.

The clonality of MMTV-induced tumors and the long delay in tumor appearance are compatible with the following model of tumor induction. Many cells are infected by MMTV, resulting in integration at numerous different sites in the infected cell population. Only if the proviral genome integrates – by chance – next to a cellular oncogene does the cell get transformed and a clonal tumor grows out. This model gained enormous strength with the discovery that tumors induced by the avian leukosis virus have the proviral genome situated next to a previously identified cellular oncogene, *c-myc* (PAYNE et al. 1982; HAYWARD et al. 1981).

Acquisition of integrated proviruses in mammary tumors was initially studied and documented by reassociation kinetics (MICHALIDES et al. 1976; MORRIS et al. 1977). Close inspection of the configuration of the integrated proviruses was made possible by the Southern transfer technique, in combination with specific probes for the MMTV genome (SHANK et al. 1978; COHEN et al. 1979b). Integrated proviruses in tumors and in infected cell lines are colinear with the viral RNA and are bordered by long terminal repeats. All of the extra proviruses that were examined closely appeared to be complete; gross deletions of the viral genome were not observed. Very informative was the discovery that endogenous and exogenous variants of MMTV can be discriminated according to their restriction maps (COHEN et al. 1979b). The appearance of restriction fragments belonging to exogenous MMTV in the DNA of mammary glands of virus-infected mice showed that integration of acquired proviruses has already occurred in nonmalignant cells. The cells are not clonal for the integration site of the provirus; characteristic viral-DNA–cell-DNA junction fragments could not be detected. The tumors that developed in virus-exposed mice contain the same acquired proviral fragments, but, in contrast to the infected mammary gland cells, tumor cells are clonal, with respect to the integration site of the provirus (COHEN et al. 1979b).

Unique junction fragments are most conveniently studied by the enzyme *Eco*RI, which cuts once in the proviral genome, generating fragments whose length is dependent on the integration site. The number of integrated proviruses can be as high as a dozen and most tumors contain at least one

(COHEN et al. 1979b; GRONER and HYNES 1980; COHEN and VARMUS 1980; FANNING et al. 1980; MICHALIDES et al. 1981b; MORRIS et al. 1980, ALTROCK et al. 1982). Acquisition of MMTV proviruses in tumors remained uncertain in only a few cases; some tumors in GR mice did not show novel viral fragments (COHEN and VARMUS 1980; MICHALIDES et al. 1981b). One explanation could be the apparent polyclonal character of these tumors, but, alternatively, one can ask whether MMTV reintegration in this strain is required for tumor formation. The mere expression of that endogenous variant or nearby genes might be sufficient (MICHALIDES et al. 1982a).

These cases remain exceptions, and as a rule, one can state that reintegration of MMTV proviruses is associated with a clonal outgrowth of cells from a mass-infected population.

Reintegration of MMTV proviruses in tumors is not restricted to mammary tumors: certain T cell leukemias in mouse strains with virulent MMTV also have amplified integrated proviruses (MICHALIDES et al. 1982b). This suggests that MMTV is somehow causally related to the induction of this disease, possible by activation of cellular oncogenes in a way discussed below. Alternatively, an event other than the integration of an MMTV provirus could be initiating transformation. This would result in a clonal appearance of a previously integrated provirus, without a role for transformation of that provirus. Needless to say, the same alternatives are valid for the integration of proviruses in mammary tumors, as well. Not withstanding these alternatives, the clonality of mammary tumors is compatible with a model in which selection occurs for cells with a provirus integrated at crucial sites, i.e., cellular oncogenes. Assuming that only a few, or even a single, oncogene would exist for mammary tumors, one might predict that individual tumors would share proviral integration sites. This did not become obvious from simple comparison of the length of viral-DNA–host-cell-DNA junction fragments in tumors, in which a complex pattern emerges due to the presence of multiple acquired proviruses over a background of multiple endogenous proviruses.

This idea of a mutational insertion of MMTV was reinforced by the discovery that another nontransforming retrovirus, avian leukosis virus, integrated near a cellular oncogene, *c-myc,* in nearly every tumor examined (HAYWARD et al. 1981; PAYNE et al. 1982). Integration leads, then, to enhanced transcription of the *c-myc* oncogene.

An analogous model for MMTV-induced tumors – integration in a specific domain in the host cell – was investigated by molecular cloning of a virus-host junction fragment from a tumor (NUSSE and VARMUS 1982). Selection of a tumor with a single acquired provirus enlarged the chance of dealing with an essential hit in the transformation. With the cellular DNA fragment as a probe, a large region surrounding the integration site was retrieved (termed MMTV *int-1*). This region turned out to be occupied by an MMTV provirus in the majority of other independent mammary tumors. The integrations were scattered over a total distance of 20 kb. This finding explains the lack of common restriction fragments in other experiments. The MMTV *int-1* region contains a domain that is transcribed in tumor cells; a 2.6-kb RNA was found in many tumors with an MMTV provirus integrated at *int-1*. Normal lactating mammary glands did not contain this RNA, suggesting, but not proving, that the transcript is a consequence of the integration. This gene is a highly likely candidate for a cellular transforming gene operating in mammary cancer, but biological evidence for its function is still lacking. It is not homologous to any of the know viral or cellular oncogenes.

Not all the tumors examined have MMTV integrated at *int-1* and the possibility arises that more such sites are present.

The mechanism of activation of the *int* gene is not clear yet. The topography of the region – integrations at both sides of the gene and at a distance up to 10 kb – precludes a promoter insertion model such as was observed for activation of *c-myc* by the ALV provirus. It is possible that the MMTV LTR carries enhancer sequences which can stimulate transcription of genes over a long distance.

MMTV *int-1* is very likely not identical with the transforming gene from mammary tumors that transforms NIH/3T3 cells, as isolated by LANE et al. (1981). This gene is not linked to MMTV sequences, even if the tumors are induced by the virus, whereas *int-1* is apparently dependent on activation in *cis* by a provirus. The relationship between these two genes, reminiscent of similar findings in chicken bursal lymphomas – a transforming gene different from *c-myc* (COOPER and NEIMAN 1981) – requires more study.


# 6 Concluding Remarks

Combining molecular biology and biology, we have begun to understand how MMTV may transform mammary gland cells. The MMTV genome itself does not appear to contain any transforming sequences, since MMTV is incapable of transforming cells in tissue culture. MMTV induces mammary tumors in mice after a long latency period, and, although the large majority of mammary gland cells become productively infected, only one or a few mammary tumors appear in an infected animal. These biological data themselves already indicate that the MMTV genes are not directly involved in the initiation of mammary tumor formation. The viral genes *gag, pol,* and *env* may contribute to some aspects of tumorigenesis, but they are at first directly involved in the replication cycle of the virion. The function of the *orf* gene in the MMTV LTR is still unknown. It is transcribed in a separate messenger RNA in mammary glands and mammary tumors of Balb/c, C3H, and GR mice (VAN OOYEN et al. 1983; WHEELER et al. 1983). In view of the structure of the MMTV genome, it has been argued that *orf* could be a transforming gene, as is the *src* gene in Rous sarcoma virus (BENTVELZEN and HILGERS 1980). However, one would expect mammary tumors to appear rapidly under the direct influence of the *orf* gene product and also to occur in untreated Balb/c mice. The *orf* gene of MMTV also does not fulfill a characteristic property of viral transforming genes, in that viral oncogenes have normal cellular homologues, whereas the MMTV LTR does not.

Conservation of the *orf* gene in proviruses of various exogenous and endogenous MMTVs suggests some function for the *orf* protein. It may be involved in the integration of MMTV proviruses in the cellular genome, it could play a role in the interaction between mammatropic hormones and the MMTV genome, or it could be involved in the processing of MMTV precursor proteins. These possibilities can be studied by generating mutations within the orf gene and seeing how these affect synthesis of viral proteins and the hormonal response. The *orf* protein seems not to be a structural MMTV protein, since antibodies prepared with disrupted MMTV virions do not detect such a protein. The search for the *orf* protein is now feasible with antibodies generated with synthetic peptides corresponding to the nucleotide sequence of the *orf* gene.

Our knowledge of the expression of endogenous MMTV proviruses is still limited. The two MMTV proviruses which are associated with mammary tumor induction loci *Mtv-1* and *Mtv-2* are clearly expressed. Their expression is tissue-specific, with mammary glands as a specific target organ. The expression of most of the other endogenous MMTV proviruses appears suppressed. This suppression is relieved by carcinogens, which may act as mutagens thereby inactivating an MMTV repressor. These limited findings suggest that positioning of the MMTV proviruses is essential to their expression. They may be located at transcriptionally inactive regions of the mouse genome, or, in case of the *Mtv-1-* and *Mtv-2*-associated MMTV proviruses, at regions which are affected by mammatropic hormones.

Some of the endogenous MMTV proviruses may also be incomplete. The endogenous MMTV provirus of *Mtv-8,* for example, appears mutated, since transfection of this proviral DNA into mink cells results in production of viral RNA and viral proteins, without release of complete virus particles (OWEN and DIGGELMAN 1983).

The lack of a direct correlation between MMTV expression and mammary tumor formation in general provides another argument for MMTV not being directly involved in mammary tumorigenesis. Some nonvirally induced mammary tumors contain enhanced levels of MMTV RNA, but this may well be a secondary effect of the tumor-inducing agent.

Based on our present understanding of retroviruses, the oncogenicity of MMTV is best explained by the insertional mutagenesis model. This model proposes that chance integration of proviral DNA in the vicinity of a crucial cellular gene results in the expression of that gene. The integration may lead to an enhanced expression or an expression in a mutated form. Expression of this crucial cellular gene results in tumor formation. This crucial cellular gene then functions as a cellular oncogene.

Integration of MMTV proviruses in mammary gland DNA takes place at multiple, possibly random, sites. However, MMTV-induced mammary tumors appear clonal, with respect to their extra MMTV proviral DNA sequences. This argues that clonal selection of particular MMTV-infected cells had occurred. Moreover, NUSSE and VARMUS (1982) found that the majority of MMTV-induced mammary tumors contained one extra provirus integrated at a common region, termed *int-1,* and that this integration leads to the expression of a 2.6 kb *int-1* mRNA.

Mammary tumor induction by MMTV, therefore, fulfills the requirements of the insertional mutagenesis model. This is also in line with biological findings. Mammary tumors appear rather late after infection and only one or a few tumors develop in an animal. Numerous integration events are apparently required, which take place in an expanding population of mammary gland cells during hormonal stimulation. Then there is still a small chance of integrating within the *int-1* region, since only one or a few mammary tumors appear.

The activation of the *int-1* gene is not likely to occur by promoter insertion, by which the LTR of the integrated provirus provides the promoter for transcription of a juxtaposed gene. The cellular *myc* oncogene in chicken is usually activated in this way in bursal lymphomas by the integration of proviral DNA of avian leukosis virus in its direct surrounding. The integrations of extra MMTV proviruses over a 20-kb region of *int-1* render its activation by promoter insertion rather unlikely. Integration of MMTV proviruses may, however, introduce enhancer sequences, present in the LTR, which exert their enhancing activity over long distances.

One also has to consider the relationship between the *int-1* gene and the transforming gene found in mammary tumors by DNA transfection assays (LANE et al. 1981). These genes are probably not identical. Since tumor formation is a multistep process, these two genes may each be involved in a different step of the tumor development and more of these genes are still to be discovered. These studies will also answer the question of whether different types of mammary tumors originate by transformation of mammary gland cells in the various stages of mammary gland differentiation, or whether the integration of extra proviruses in different crucial regions of the cellular genome results in different tumor types. In the latter case, one would expect a variety of oncogenes to be involved in mammary tumor development. The insertional mutagenesis model for MMTV-induced mammary tumor formation offers new avenues to mammary cancer research. It shows that MMTV functions as a trigger for tumor formation and that the essential genes for tumor development are part of the normal genetic makeup of the cells. Knowledge of these genes, their products, and their action will be the next step in the MMTV-induced mammary cancer research.

*Acknowledgements.* A VAN OOYEN is supported by the Dutch Queen Wilhelmina Foundation. We thank H.E. VARMUS and J. MAJORS for communication of results prior to publication and N. VAN NULAND for preparation of the manuscript.

# References

Altrock BW, Cardiff RD, Puma PP, Lund JK (1982) Detection of acquired provirus sequences in mammary tumors from low-expressive, low-risk mice. J Natl Cancer Inst 68:1037–1041

Axel R, Schlom J, Spiegelman S (1972) Evidence for translation of viral-specific RNA in cells of a mouse mammary carcinoma. Proc Natl Acad Sci USA 69:535–538

Bentvelzen P, Hilgers J (1980) The murine mammary tumor virus. In: Klein G (ed) Viral oncology. Raven, New York, pp 311–355

Bird AP, Southern EM (1978) Use of restriction enzymes to study eukaryotic DNA methylation. I. The methylation pattern in ribosomal DNA from *Xenopus laevis*. J Mol Biol 118:27–47

Bittner JJ (1936) Some possible effects of nursing on the mammary gland tumor incidence in mice. Science 84:162–169

Bishop JM (1978) Retroviruses. In: Snell EE (ed) Annual review of biochemistry, Vol. 47. Annual Reviews Inc, Palo Alto, pp 35–89

Breznik T, Cohen JG (1982) Altered methylation of endogenous viral promoter sequences during mammary carcinogenesis. Nature 295:255–257

Buetti E, Diggelman H (1981) Cloned mouse mammary tumor virus DNA is biologically active in transfected mouse cells and its expression is stimulated by glucocorticoid hormones. Cell 23:335–345

Butel JS, Dusing-Swartz S, Socher SH, Medina D (1981) Partial expression of endogenous mouse mammary tumor virus in mammary tumors induced in Balb/c mice by chemical, hormonal and physical agents. J Virol 38:571–580

Callahan R, Drohan W, Gallahan D, Hoostelaere L, Potter M (1982) Novel class of mouse mammary tumor virus-related DNA sequences found in all species of *Mus*, including mice lacking the virus proviral genome. Proc Natl Acad Sci USA 79:4113–4417

Cohen JC (1980) Methylation of milk-borne and genetically transmitted mouse mammary tumor virus proviral DNA. Cell 19:652–662

Cohen JC, Varmus HE (1979) Endogenous mammary tumor virus DNA varies among wild mice and segregates during inbreeding. Nature 278:418–423

Cohen JC, Varmus HE (1980) Proviruses of mouse mammary tumor virus in normal and neoplastic tissues from GR and C3Hf mouse strains. J Virol 35:298–305

Cohen JC, Majors JE, Varmus HE (1979a) Organization of mouse mammary tumor virus-specific DNA endogenous to Balb/c mice. J Virol 32:483–496

Cohen JC, Shank PR, Morris VL, Cardiff RD, Varmus HE (1979b) Integration of the DNA of mouse mammary tumor virus in virus-infected normal and neoplastic tissue of the mouse. Cell 16:333–345

Cooper GM, Neiman P (1981) Two distinct candidate-transforming genes of lymphoid leukosis virus-induced neoplasms. Nature 292:857–858

Dickson C, Peters G (1981) Protein-coding potential of mouse mammary tumor virus genome RNA as examined by in vitro translation. J Virol 37:36–47

Dickson C, Smith R, Peters G (1981) In vitro synthesis of polypeptides encoded by the long terminal repeat region of mouse mammary tumor virus DNA. Nature 291:511–513

Diggelman H, Vessaz AL, Buetti E (1982) Cloned endogenous mouse mammary tumor virus DNA is biologically active in transfected mouse cells and its expression is stimulated by glucocorticoid hormones. Virol 122:332–341

Dion AS, Heine J, Pomenti AA, Korb J, Weber GH (1977) Electrophoretic analysis of the molecular weight of murine mammary tumor virus RNA. J Virol 22:822–825

Donehower LA, Huang AL, Hager GL (1981) Regulatory and coding potential of the mouse mammary tumor virus long terminal redundancy. J Virol 37:226–238

Drohan WN, Benade LE, Graham DE, Smith EH (1982) Mouse mammary tumor virus proviral sequences congenital to C3H/Sm mice are differentially hypomethylated in chemically induced, virus-induced and spontaneous mammary tumors. J Virol 43:876–884

Dudley JP, Varmus HE (1981) Purification and translation of murine mammary tumor virus mRNAs. J Virol 39:207–218

Dudley JP, Butel JS (1979) Effect of dexamethasone on expression of endogenous mouse tumor virus sequences in Balb/c tumor cell lines. Virology 96:453–462

Dudley JP, Rosen JM, Butel JS (1978a) Differential expression of poly(A)-adjacent sequences of mammary tumor virus RNA in murine mammary cells. Proc Natl Acad Sci USA 75:5797–5801

Dudley JP, Butel JS, Socher SH, Rosen JM (1968b) Detection of mouse mammary tumor virus RNA in Balb/c tumor cell lines of non-viral etiologies. J Virol 28:743–752

Duesberg PH, Cardiff RD (1968) Structural relationships between the RNA of mammary tumor virus and those of other RNA tumor viruses. Virology 36:696–700

Dusing-Swartz S, Medina D, Butel JS, Socher SH (1979) Mouse mammary tumor virus genome expression in chemical carcinogen-induced mammary tumors in low- and high-tumor-incidence mouse strains. Proc Natl Acad Sci USA 76:5360–5364

Edmonds M, Caramela MG (1969) The isolation and characterization of adenosine-monophosphate-rich polynucleotides synthesized in Ehrlich ascites cells. J Biol Chem 244:1314–1324

Fanning TG, Puma JP, Cardiff R (1980) Selective amplification of mouse mammary tumor virus in mammary tumors of GR mice. J Virol 36:109–114

Fanning TG, Vassos AB, Cardiff RD (1982) Methylation and amplification of mouse mammary tumor virus DNA in normal, premalignant and malignant cells of GR mice. J Virol 41:1007–1013

Fasel NK, Pearson EK, Buetti E, Diggelman H (1982) The region of mouse mammary tumor virus DNA containing the long terminal repeat includes a long coding sequence and signals for hormonally regulated transcription. EMBO J 1:3–7

Foulds L (1975) Mammary neoplasia in laboratory animals. In: L. Foulds (Ed). Neoplastic Development, Academic Press, London, 457–548

Gilboa E, Mitra SW, Goff S, Baltimore D (1979) A detailed model of reverse transcription and tests of crucial aspects. Cell 18:93–100

Groner B, Hynes NE (1980) Number and location of mouse mammary tumor virus proviral DNA in mouse DNA of normal tissue and of mammary tumors. J Virol 33:1013–1025

Groner B, Buetti E, Diggelman H, Hynes NE (1980) Chracterization of endogenous and exogenous mouse mammary tumor virus proviral DNA with site-specific molecular clones. J Virol 36:734–745

Hayward WG, Neel BE, Astrin SM (1981) Activation of a cellular oncogene by promoter insertion in ALV-induced lymphoid leukosis. Nature 290:475–480

Hilgers J, Sluyser M (eds) (1981) Mammary tumors in the mouse. Elsevier/North-Holland, Amsterdam

Hynes NE, Kennedy N, Rahmsdorf U, Groner B (1981) Hormone-responsive expression of an endogenous proviral gene of mouse mammary tumor virus after molecular cloning and gene transfer into cultures cells. Proc Natl Acad Sci USA 78:2038–2042

Imai S, Morimoto J, Tsubura Y, Hilgers J (1982) Mammary tumor induction in inbred mouse strains with urethane is not accompanied by changes in expression of B- and C-type retroviral structural proteins. Int J Cancer 30:101–106

Kennedy N, Knedlitschek G, Groner B, Hynes NE, Herrlich P, Michalides R, van Ooyen AJJ (1982) The long terminal repeats of an endogenous mouse mammary tumor virus are identical and contain a long open reading frame extending into adjacent sequences. Nature 295:622–624

Klemenz R, Reinhardt M, Diggelman H (1981) Sequence determination of the 3′ end of mouse mammary tumor virus RNA. Mol Biol Rep 7:123–126

Lane MA, Sainten A, Cooper GM (1981) Activation of related transforming genes in mouse and human mammary carcinomas. Proc Natl Acad Sci USA 78:5185–5189

Lim L, Canellakis ES (1970) Adenine-rich sequences associated with rabbit reticulocyte messenger RNA. Nature 227:710–712

Macinnes JI, Lee-Chan ECM, Percy DH, Morris VL (1981) Mammary tumors from GR mice contain more than one population of mouse mammary tumor virus-infected cells. Virology 113:119–129

Majors JE, Varmus HE (1981) Nucleotide sequences at host-proviral junctions for mouse mammary tumor virus. Nature 289:253–258

Mandel JL, Chambon P (1979) DNA methylation: organ-specific variations in the methylation pattern within and around ovalbumin and other chicken genes. Nucleic Acids Res 7:2081–2103

McGrath CM, Marineau EJ, Voyles BA (1978) Changes in MuMTV DNA and RNA levels in Balb/c mammary epithelial cells during malignant transformation by exogenous MuMTV and by hormones. Virology 87:339–353

McGrath CM, Prass WA, Maloney TM, Jones RF (1981) Induction of endogenous mammary tumor virus expression during hormonal induction of mammary adenoacarcinomas and carcinomas of Balb/c female mice. J Natl Cancer Inst 67:841–852

Medina D (1978) Preneoplasia in breast cancer. In: McGuire W (ed) Breast cancer, vol. 2. Plenum, New York, pp 47–102

Michalides R, Nusse R (1981) Molecular biology of the mouse mammary tumor virus. In: Hilgers J, Sluyser M (eds) Mammary tumors in the mouse. Elsevier/North-Holland, Amsterdam, pp 465–503

Michalides R, Schlom J (1975) Relationship in nucleic acid sequences between mouse mammary tumor virus variants. Proc Natl Acad Sci USA 72:4635–4639

Michalides R, Valahakis G, Schlom J (1976) A biochemical approach to the study of the transmission of mouse MTV in mouse strains RIII and C3H. Int J Cancer 18:105–115

Michalides R, van Deemter L, Nusse R, Röpcke G, Boot L (1978) Involvement of mouse mammary tumor virus in spontaneous and hormone-induced mammary tumors in low-mammary-tumor mouse strains. J Virol 27:551–559

Michalides R, van Deemter L, Nusse R, Hageman P (1979) Induction of mouse mammary tumor virus RNA in mammary tumors of Balb/c mice treated with urethane, X-irradiation and hormones. J Virol 31:63–72

Michalides R, Van Nie B, Hynes NE, Groner B (1981a) Mammary tumor induction loci in GR and DBAf mice contain one provirus of the mouse mammary tumor virus. Cell 23:165–173

Michalides R, Wagenaar E, Groner B, Hynes NE (1981b) Mammary tumor virus proviral DNA in normal murine tissue and nonvirally induced mammary tumors. J Virol 39:367–376

Michalides R, Wagenaar E, Sluyser M (1982a) Mammary tumor virus DNA as a marker for genotypic variance within hormone-responsive GR mammary tumors. Cancer Res 42:1154–1158

Michalides R, Wagenaar E, Hilkens J, Hilgers J, Groner B, Hynes NE (1982b) Acquisition of proviral DNA of mouse mammary tumor virus in thymic leukemia cells from GR mice. J Virol 43:819–829

Morris VL, Medeiros E, Ringold GM, Bishop JM, Varmus HE (1977) Comparison of mouse mammary tumor virus-specific DNA in inbred, wild and Asian mice, and in tumors and normal organs from inbred mice. J Mol Biol 114:73–91

Morris VL, Vlasschaert JE, Beard CL, Milazzo MF, and Bradbury WC (1980) Mammary tumors from Balb/c mice with a reported high mammary tumor incidence have acquired new mammary tumor virus DNA sequences Virology 100, 101–109

Nusse R, Michalides R, Röpcke G, Boot LM (1980) Quantification of mouse mammary tumor virus structural proteins in hormone-induced mammary tumors of low-mammary-tumor mouse strains. Int J Cancer 25:377–383

Nusse R, Varmus HE (1982) Many tumors induced by the mouse mammary tumor virus contain a provirus integrated in the same region of the host genome. Cell 31:99–109

Owen P, Diggelman H (1983) Cloned mouse mammary tumor virus DNA exhibits glucocorticoid-dependent expression in simian virus 40-transformed mink cells. J Virol 45:148–154

Pauley RJ, Medina D, Socher SH (1979) Hormonal regulation of murine mammary tumor virus RNA expression during mammary tumorigenesis in Balb/c mice. J Virol 32:557–566

Payne GS, Bishop JM, Varmus HE (1982) Multiple arrangements of viral DNA and an activated host oncogene in bursal lymphomas. Nature 295:209–214

Peters G, Glover C (1980) tRNAs and priming of RNA-directed DNA synthesis in mouse mammary tumor virus. J Virol 35:31–40

Ringold GM, Cardiff RD, Varmus HE, Yamamoto KR (1977) Infection of cultured hepatoma cells by mouse mammary tumor virus. Cell 10:11–18

Robertson DL, Varmus HE (1981) Dexamethasone induction of the intracellular RNAs of mouse mammary tumor virus. J Virol 40:673–682

Schlom J, Colcher D, Spiegelman S, Gillespie S, Gillespie D (1973a) Quantitation of RNA tumor viruses and viruslike particles in human milk by hybridization to polyadenylic acid sequences. Science 179:696–698

Schlom J, Michalides R, Hehlmann R, Spiegelman S, Bentvelzen P, Hageman P (1973b) A comparative study of the biologic and molecular basis of murine mammary carcinoma; a model for human breast cancer. J Natl Cancer Inst 51:541–551

Shank PR, Cohen JC, Varmus HE, Yamamoto KR, Ringold GM (1978) Mapping of linear and circular forms of mouse mammary tumor virus DNA with restriction endonucleases:

evidence for a large specific deletion occurring at high frequency during circularization. Proc Natl Acad Sci USA 75:2112–2116

Sluyser M, Evers SG, de Goey CCJ (1976) Sex hormone receptors in mammary tumors of GR mice. Nature 263:386–389

Smith GH, Arthur LA, Medina D (1980) Evidence of separate pathways for viral and chemical carcinogenesis in C3H/StWi mouse mammary glands. Int J Cancer 26:373–379

Teramoto YA, Medina DA, McGrath C, Schlom J (1980) Noncoordinate expression of murine mammary tumor virus gene products. Virology 107:345–353

Traina VL, Taylor BA, Cohen JG (1981) Genetic mapping of endogenous mouse mammary tumor viruses: Locus characterisation, segregation and chromosomal distribution. J Virol 40:735–744

Vaidya AB, Lasfargues EY, Heubel G, Lasfargues JC, Moore DH (1976) Murine mammary tumor virus: characterization of infection of non-murine cells. J Virol 18:911–917

Van Nie R (1981) Mammary tumorigenesis in the GR mouse strain. In: Hilgers J, Sluyser M (eds) Mammary tumors in the mouse. Elsevier/North-Holland, Amsterdam, pp 201–266

Van Nie R, Moes de J (1977) Development of a congeneic line of the GR mouse strain without early mammary tumors. Int J Cancer 20:588–594

Van Nie R, Verstraeten AA (1975) Studies of genetic transmission of mammary tumor virus by C3Hf mice. Int J Cancer 16:922–931

Van Nie R, Verstraeten AA, de Moes J (1977) Genetic transmission of mammary tumor virus by GR mice. Int J Cancer 19:383–390

Van Ooyen AJJ, Michalides R, Nusse R (1983) Structural analysis of a 1.7-kb mouse mammary tumor virus-specific RNA. J Virol 46:362–370

Varmus HE, Bishop JM, Nowinski RC, Sarkar NH (1972) Mammary tumor virus-specific nucleotide sequences in mouse DNA. Nature New Biol 238:189–191

Varmus HE, Quintrell N, Medeiros E, Bishop JM, Nowinsky RC, Sarkar NH (1973) Transcription of mouse mammary tumor virus genes in tissues from high- and low-incidence mouse strains

Waters L (1978) Lysine tRNA is the predominant tRNA in murine mammary tumor virus. Biochem Biophys Res Commun 81:822–827

Weiss R, Teich N, Varmus H (eds) (1982) RNA tumor viruses. Cold spring Harbor monograph series 10C. Molecular biology of tumor viruses, 2nd edn. Cold Spring Harbor Laboratory

Wheeler DA, Butel JS, Medina D, Cardiff RD, Hager GL (1983) Transcription of mouse mammary tumor virus: identification of a candidate mRNA for the LTR gene product. J Virol 46:42–49

# Regulation of Mouse Mammary Tumor Virus Gene Expression by Glucocorticoid Hormones

GORDON M. RINGOLD

## 1 Introduction

Steroid hormones play an important role in regulating gene expression in metazoan organisms. A large body of work indicates that the various classes of steroids function via a unified "two-step" mechanism, originally pro-

Department of Pharmacology, Stanford University, School of Medicine, Stanford, CA 94305, USA

posed by JENSEN et al. (1968) and GORSKI et al. (1968). Simply stated, the hormone first binds to a soluble receptor protein, which is usually found in the cytoplasm, and induces a structural alteration that increases the receptor's affinity for DNA. This "activated" form of the steroid-receptor complex accumulates within the nucleus of the cell, where, by unknown mechanisms, it stimulates (and, perhaps in some instances, inhibits) the transcription of specific genes. The classes of new messenger RNAs produced in response to a particular steroid are, in large part, cell or tissue specific and their utilization in production of new proteins leads to the characteristic hormonal response of the target cell. The primary role of the steroid molecule could thus be envisioned as that of an allosteric effector that unmasks a DNA-binding site on the receptor protein. The details of the molecular events involved in this activation process and the mechanisms by which the steroid-receptor complex stimulates specific gene expression remain obscure. For more extensive discussion of the basic two-step model of steroid hormone action and the evidence in support of it, the reader is directed to one or more of the many reviews on this subject (GORSKI and GANNON 1976; YAMAMOTO and ALBERTS 1976; HIGGINS and GEHRING 1978).

There are several aspects of the model described above that raise fundamental questions about eukaryotic gene regulation. How does the steroid-receptor complex regulate specific gene transcription and by what means does it find those genes within the nucleus? Are there specific and high-affinity binding sites associated with regulatory regions of the target genes? What aspects of the transcription machinery are affected by the presence of the steroid-receptor complex? What effect, if any, might the hormone have on the chromatin configuration of the regulated gene? These and other questions are central not only to our understanding of steroid hormone action but to the broader issue of gene regulation during development and differentiation. In this review, I will focus on the regulation of mouse mammary tumor virus (MMTV) gene expression by glucocorticoid hormones as a useful model system for approaching these questions.

## 2 Mouse Mammary Tumor Virus

As is true of all retroviruses, MMTV is a single-stranded RNA virus (DUESBERG and CARDIFF 1968) containing two identical subunits of genomic RNA (approximately 9000 bases in length) and several associated small RNAs, including a hydrogen-bonded tRNA lys (PETERS and GLOVER 1980) used as a primer for viral DNA synthesis. The virion itself is composed of a nucleoprotein core containing the RNA, several nonglycosylated structural proteins, and a polymerase, the so-called reverse transcriptase, that transcribes the viral genomic RNA into double-stranded DNA. Surrounding the core is an envelope containing components of the plasma membrane of the infected cell and the major viral glycoprotein (gp 52). For a more

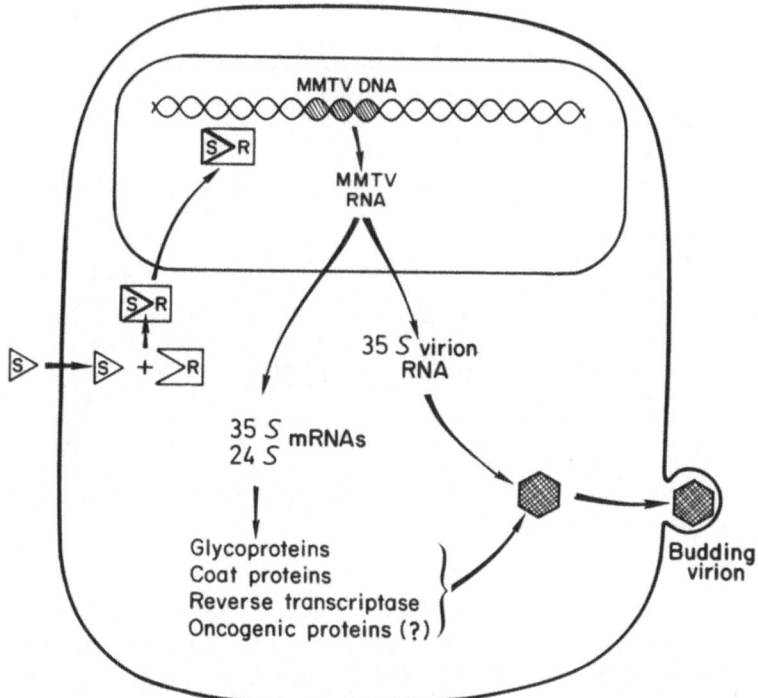

**Fig. 1.** Stimulation of MMTV production by glucocorticoid hormones. The steroid molecule enters the cell, apparently by a passive mechanism, and binds tightly to a soluble, cytostolic receptor protein. The steroid-receptor complex undergoes modification to an "activated" state that leads to its accumulation within the nucleus. It is presumed that binding to DNA, and perhaps specifically to MMTV DNA, leads to an increase in the rate of synthesis of viral RNA. Several RNA species, including 35-*S* genomic RNA and 35-*S* and 24-*S* mRNAs, can be found in infected cells; there are indications that additional smaller species of viral RNA may also be produced. Viral proteins and genomic RNA associate to form the core of the virus particle which then buds from the plasma membrane

complete description of MMTV proteins, their synthesis, and assembly, the reader is directed to the review by DICKSON and PETERS in this volume.

During infection of a cell, a linear double-stranded DNA copy of the viral RNA is synthesized by the viral reverse transcriptase (VAIDYA et al. 1976; RINGOLD et al. 1977c; RINGOLD et al. 1978). Sometime later, MMTV DNA becomes covalently linked to the host DNA, where the resultant provirus resides as a stable genetic element in the progeny of the infected cell. A likely intermediate in the integration event is a covalently closed circular form of MMTV DNA that can be detected under appropriate experimental conditions in the nuclei of infected cells (RINGOLD et al. 1977c, 1978). Once formed, the provirus is transcribed by the cellular RNA polymerase II (STALLCUP et al. 1978) forming several species of RNA, including genomic-sized RNA for encapsidation into progeny virus, as well as for translation of the coat proteins and polymerase (SEN et al. 1979). In addition,

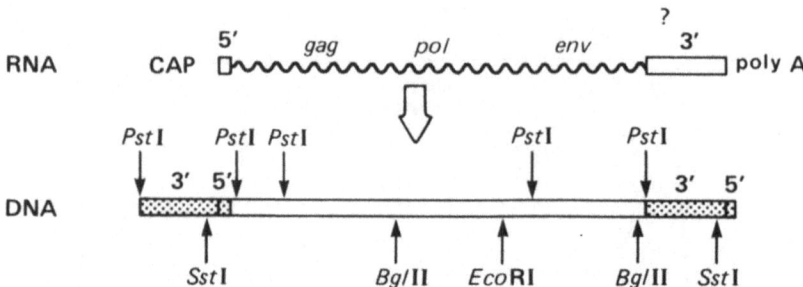

**Fig. 2.** The structure of MMTV RNA and DNA. The intact viral RNA (9000 bases) is illustrated with its coding regions indicated: *gag*, group-specific antigens, core proteins; *pol*, viral reverse transcriptase; *env*, envelope glycoproteins; *?*, a postulated protein encoded within a 3′ open reading frame. The linear form of MMTV DNA contains long terminally repeated (LTR) sequences derived from both the 5′ and 3′ ends of viral RNA; these are shown as the *stippled regions* designated 3′ (1200 bp) and 5′ (130 bp). Upon integration into chromosomal DNA, this structure is maintained with minor modifications; two nucleotides of viral DNA are lost from each end and a 6-bp duplication of the cell DNA, arising from the integrative recombination, flanks the proviral DNA

smaller spliced RNAs are synthesized as templates for translation of the envelope glycoproteins and perhaps a protein from an open reading frame present at the 3′ end of MMTV RNA (ROBERTSON and VARMUS 1979; SEN et al. 1979; GRONER et al. 1979; DUDLEY and VARMUS 1981; DICKSON and PETERS 1981; DICKSON et al. 1981). It is the production of MMTV RNAs from proviral DNA that is under glucocorticoid control (Fig. 1).

## 2.1  Structure of Mouse Mammary Tumor Virus DNA

Before proceeding with the discussion of steroid effects on MMTV, a short digression is required to discuss the novel structure of MMTV DNA. As for other retroviruses, the synthesis of proviral DNA by reverse transcriptase is a fascinating and remarkably complex process (GILBOA et al. 1979; for review see TEMIN 1981 and VARMUS 1982). An important feature of such DNAs is that they are longer than the parental RNA; this extra length arises by duplication of sequences present at the extreme 5′ and 3′ ends of the viral RNA. These duplications are present at the ends of the linear viral DNA and are thus commonly referred to as the "long terminal repeats" or LTRs (VARMUS 1982). In the case of MMTV DNA, the LTRs are approximately 1350 base pairs in length, with about 130 of these arising from the 5′ end of viral RNA (Fig. 2). The LTR sequences from several isolates of MMTV DNA have been determined (DONEHOWER et al. 1981; KENNEDY et al. 1982; FASEL et al. 1982). From the sequence analysis of LTRs, host DNA-proviral junctions, and MMTV RNA itself (MAJORS and VARMUS 1981; DONEHOWER et al. 1981; KLEMENZ et al. 1981), several points can be made:

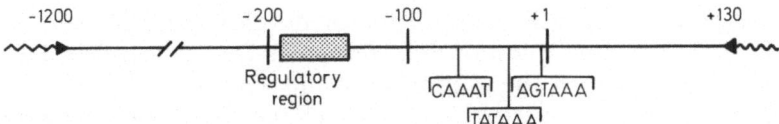

**Fig. 3.** Points of interest on MMTV LTR. Sequences implicated in initiation of transcription (TATAAA) and (CAAAT) reside approximately 25 and 75 nucleotides respectively, upstream from the start of transcription. The sequence AGTAAA, which may serve as a signal for polyadenylation of MMTV RNA in the right LTR, is located in a position corresponding to 15–20 nucleotides from the 3′ end of viral RNA. A 6-bp inverted repeat (◀, ▶) flanks the ends of each LTR. The glucocorticoid regulatory region (*stippled*) maps between 140 and 190 residues upstream from the start of transcription. This figure respresents the left LTR of an integrated viral genome with cellular DNA (ww) and the remainder of the viral DNA (ww) indicated

1. Six nucleotides adjacent to the LTR are directly repeated in host DNA, presumably as a consequence of the integrative recombination.
2. The sequence, TATAAA, which is thought to be a signal for initiation by RNA polymerase II (GOLDBERG 1978), is found 25–30 nucleotides upstream from the viral RNA cap site.
3. A signal for polyadenylation of MMTV RNA (AGTAAA) (PROUDFOOT and BROWNLEE 1974) is located 17 bases upstream from the known 3′ end of viral RNA.

Based on these observations and direct experimentation (see below), it is likely that viral RNA synthesis initiates within the left-hand LTR and is terminated by a processing event at a site derived from the right-hand LTR. Figures 2 and 3 summarize several structural features of MMTV RNA and DNA that will be useful in the following discussion.

# 3 Hormonal Regulation of Mouse Mammary Tumor Virus Gene Expression

## 3.1 Mammary Tumors and Hormones

Early observations of mammary tumor development in mice implicated a variety of hormones in both the incidence and progression of tumors (NANDI and MCGRATH 1973). In addition to the obvious sex dependence of the tumors, tumor frequency and growth rate are often highest during pregnancy. A variety of hormonal stimuli, including steroid and polypeptide hormones, have marked effects on the incidence of mammary tumors in many mouse strains, and administration of estrogens to male mice may lead to a high incidence of mammary tumorigenesis (BERN and NANDI 1961). The role of adrenocortical hormones in tumor incidence is probably minor. However, since glucocorticoids are essential for mammary development, they may play a permissive role in tumor development (TOPPER and FREEMAN 1980). There is no question that receptors for glucocorticoids and sex steroids are present in the mammary gland (SHYAMALA 1973; RINGOLD et al. 1975b); however, it is unclear whether the interaction between any of the hormones and MMTV is in any way related to mammary tumorigenesis.

## 3.2 Hormonal Stimulation of Mouse Mammary Tumor Virus Production

Treatment of mice or primary explants of mammary tumors with glucocorticoids results in the accumulation of MMTV particles within the tumor tissue (SMOLLER et al. 1961). Successful culturing of tumor explants that produce MMTV allowed McGRATH (1971) to demonstrate directly that hydrocortisone and insulin stimulate MMTV production. The role of insulin remains unclear, however, since several reports demonstrate no effect of this hormone (DICKSON et al. 1974; CARDIFF et al. 1976). Using a novel culture system in which mammary tumor cells are grown on a floating collagen matrix, YANG et al. (1977) have reported that MMTV production is enhanced by treatment with prolactin, growth hormone, or placental lactogen. This effect was additive to that of glucocorticoids. However, under standard growth conditions, only glucocorticoids (e.g., hydrocortisone and dexamethasone) have a reproducible effect on MMTV production by cells in culture (FINE et al. 1974; DICKSON et al. 1974; PARKS et al. 1974; RINGOLD et al. 1975a, b). In general, 10–20 times more virus is produced by glucocorticoid-treated mouse mammary tumor cells than by control cultures. The possible evolutionary significance of this phenomenon, to either the virus or the host, remains obscure. As will be discussed extensively below, this hormonal effect is retained in heterologous cells infected with MMTV.

## 4 Glucocorticoid Induction of Mouse Mammary Tumor Virus RNA Is a Primary Response

Initial characterization of the glucocorticoid effect on MMTV production revealed a rapid increase in the steady-state level of intracellular viral RNA (PARKS et al. 1974; RINGOLD et al. 1975a, b; SCOLNICK et al. 1976). A significant problem in studying the initial steps in steroid hormone action has been to establish that the response under investigation represents a direct or primary action of the steroid-receptor complex at the transcriptional level.

Several types of experiments strongly support the contention that the glucocorticoid induction of MMTV RNA is indeed a primary hormonal response. First, complete induction of MMTV RNA occurs in the absence of protein synthesis but is blocked by inhibitors of RNA synthesis, such as actinomycin D (RINGOLD et al. 1975b; SCOLNICK et al. 1976). Thus, it seems likely that the induction of viral RNA does not require prior production of a protein intermediate. This criterion for a primary response is best exemplified by the induction of puffs in the polytene chromosomes of DROSOPHILA by the moulting hormone, ecdysone. In this case, a set of so-called "early" puffs are induced by ecdysone in the absence of protein synthesis, whereas the "middle" and "late" puffs are not (ASHBURNER et al. 1973). In the extensively studied case of ovalbumin mRNA induction by estrogen, it is noteworthy that inhibitors of protein synthesis abolish induction (MCKNIGHT 1978).

Further evidence that the effect of glucocorticoids on MMTV is direct is afforded by studies on the rate of synthesis of MMTV RNA. Initial experiments were performed by measuring the proportion of MMTV RNA contained in total pulse-labeled RNA from mammary tumor cells and rat hepatoma cells infected with MMTV. The results of these studies indicate that an increase in the rate of transcription of proviral DNA can account for most, if not all, of the increase in steady-state levels of MMTV RNA (RINGOLD et al. 1977b; YOUNG et al. 1977). Moreover, the induced rate of viral RNA synthesis is achieved within 15 min after addition of the hormone. More recent experiments, using the ability of endogenous RNA polymerase to elongate nascent RNA strands in isolated nuclei, corroborate these conclusions and demonstrate that an increase in viral RNA synthesis can be detected within 5 min after addition of hormone (STALLCUP et al. 1978; UCKER et al. 1981).

A large body of work, primarily in the chick oviduct system (for reviews see SCHIMKE et al. 1975; O'MALLEY et al. 1979), led to the suggestion that steroid-receptor complexes act by altering the rate of transcription of specific genes. It is of interest, however, that the demonstration that dexamethasone stimulates the rate of synthesis of MMTV RNA provided the first definitive experimental evidence that steroid hormones may act via such a mechanism.

## 4.1 Role of the Receptor in Mouse Mammary Tumor Virus Induction

As summarized in the introduction, the binding of a steroid hormone to its cognate receptor appears to elicit an allosteric modification that results in nuclear binding of the hormone-receptor complex. Several experimental observations indicate that this view of steroid action is applicable to the induction of MMTV RNA. First, extracts of mammary tumor cells (as well as responsive heterologous cells) contain specific glucocorticoid receptors (RINGOLD et al. 1975b; YOUNG et al. 1975; SHYAMALA and DICKSON 1976). These receptors bind to DNA cellulose under appropriate conditions and migrate to the nucleus in a temperature-dependent fashion when occupied by active hormones. Half-maximal induction of MMTV RNA occurs

at concentrations of dexamethasone that half-saturate receptors, and progesterone, a competitive antagonist at the glucocorticoid receptor, blocks the
induction of MMTV RNA by dexamethasone (RINGOLD et al. 1975b;
YOUNG et al. 1975). Lastly, genetic evidence (see below) suggests that
MMTV is no longer inducible in cells that lack functional glucocorticoid
receptors (GROVE et al. 1980).

# 5   Infection of Heterologous Cells with Mouse Mammary Tumor Virus

The ability to infect heterologous tissue culture cells with MMTV was first
reported by VAIDYA et al. (1976). Since the infected cells (mink and cat)
were unaltered in growth properties or morphology, biochemical tests were
used to assess infection. Using a similar approach, rat hepatoma cells were
also shown to be susceptible to infection by MMTV (RINGOLD et al. 1977a).
With regard to steroid hormone action, the most striking result of infecting
heterologous cells was that MMTV RNA and proteins remain highly glucocorticoid inducible (VAIDYA et al. 1976; RINGOLD et al. 1977a). Since these
cells are ostensibly devoid of endogenous MMTV sequences in their genomes, the viral RNA must be synthesized from newly acquired proviruses.

The induction of MMTV RNA also appears to be due to a direct effect
of the glucocorticoid-receptor complex in heterologous cells (RINGOLD unpublished). Thus, it seems likely that the site(s) on the glucocorticoid receptor involved in recognizing MMTV sequences, in addition to the hormone
binding site, must be conserved among mammalian species.

## 5.1   Integration of Mouse Mammary Tumor Virus DNA

Restriction endonuclease mapping of the MMTV genome was performed
using small amounts of unintegrated viral DNA present in infected rat
hepatoma (HTC) cells (SHANK et al. 1978; RINGOLD et al. 1978). The structure of proviral DNA in clones of infected HTC cells and in tumors arising
by natural infection with MMTV was analyzed by the blotting and hybridization procedure of SOUTHERN (1975). Two major conclusions could be
drawn from such studies:

1. A very large number of sites in cellular DNA can be utilized for proviral
   integration (RINGOLD et al. 1979; COHEN et al. 1979; GRONER and HYNES
   1980). In fact, integration of viral DNA may be completely random.
2. The integration event always utilizes the same sites on viral DNA, thereby
   preserving the structure containing both LTRs (Fig. 2) (RINGOLD et al.
   1979; COHEN et al. 1979).

The implications of these observations for steroid responsiveness of MMTV are far-reaching. First, it makes untenable the proposition that the glucocorticoid regulation of MMTV is imparted by flanking cellular sequences. Second, since proviral DNA maintains both copies of the terminal repeat and the 5' end of viral RNA corresponds to a site within the left-hand LTR (see Fig. 2), it seems likely that the LTR itself contains a hormone-regulated promoter. This will be discussed in detail in a following section.

## 5.2 Expression of Proviral DNA in Virus-Infected Cells

Analysis of infected clones of HTC cells and individual mammary tumors revealed that not only were the sites of integration different but the absolute number of proviruses also varied widely (RINGOLD et al. 1979; COHEN et al. 1979; GRONER and HYNES 1980; MICHALIDES et al. 1981). Characterization of MMTV RNA levels in clones of infected rat hepatoma cells containing from 1 to 30 copies of viral DNA has revealed some interesting facts (RINGOLD et al. 1979):

1. There is generally more viral RNA in cells containing multiple proviruses; this correlation holds both in the absence and presence of dexamethasone.
2. The extent of induction by the hormone varies substantially and appears to be unreleated to DNA copy number.
3. Some clones that harbor only one or two proviruses make no RNA, either in the presence or absence of dexamethasone; this could be accounted for by random integration of proviruses into chromosomal sites that are transcriptionally inactive.

The influence of chromosomal position on the expression of MMTV DNA has been studied in two clones of infected HTC cells containing a single provirus (FEINSTEIN et al. 1982). One clone expresses MMTV RNA and is glucocorticoid-inducible, whereas no RNA could be detected in the other. Using the susceptibility of the provirus to digestion by DNAse I as an assay, these investigators have concluded that the transcriptionally inactive provirus integrated into a region of the genome present in a heterochromatinlike structure. This assay has been used extensively to correlate gene activity with differences in chromatin structure (WEINTRAUB and GROUDINE 1976).

These studies provide a rather compelling case for a role of chromatin structure or chromosomal position in controlling the expression of a newly introduced set of viral genes. Nevertheless, whenever there is basal expression of an MMTV provirus (or at least the capacity to be transcribed), it is glucocorticoid-inducible. Thus, it is clear that the ability of MMTV to respond to glucocorticoids is intimately associated with the viral genome itself. The chromatin-associated factors influencing the absolute level of expression of an integrated provirus remain obscure and will undoubtedly provide an exciting, but challenging, avenue of investigation.

# 6  Genetic Approaches to Glucocorticoid Action

In delineating the components of a complex regulatory system, it is often useful to select mutants with defects in one or more steps of that system. The most fruitful use of genetics in studying hormone action has been the characterization of adenylate cyclase activation and elucidation of the role of the GTP-binding protein that couples membrane receptors to the catalytic subunit of the cyclase (BOURNE et al. 1975; ROSS and GILMAN 1977). Selection of cells with altered responses to glucocorticoids was first accomplished by Tomkins and his colleagues (BAXTER et al. 1971; ROSENAU et al. 1972), using the S49 mouse lymphoma cell line. Since many mouse T cell lines die in response to glucocorticoids, a large number of glucocorticoid-resistant variants were easily isolated (SIBLEY and TOMKINS 1974a, b; PFAHL et al. 1978). Most, if not all, of the lymphoma variants lack glucocorticoid receptors or contain receptors with altered physical properties (YAMAMOTO et al. 1974).

## 6.1  Isolation of Mouse Mammary Tumor Virus-Infected Rat Hepatoma Cells with Defects in Glucocorticoid Responsiveness

MMTV-infected HTC cells have been used more recently in attempts to select mutants in both cellular and viral functions required for glucocorticoid responsiveness (GROVE et al. 1980; GROVE and RINGOLD 1981; FIRESTONE and YAMAMOTO, 1983). Since the major MMTV glycoprotein, gp 52, is expressed in a hormone-dependent manner on the surface of infected cells (RINGOLD et al. 1977a), immunological techniques have been applied to the isolation of cells that are incapable of inducing this protein.

GROVE  et al. (1980) used a fluorescence-activated cell sorter (FACS) to separate cells expressing varying amounts of gp 52; staining of cells was performed with rabbit antibody against gp 52, followed by fluorescein-conjugated goat antirabbit IgG. By multiple rounds of enrichment in the FACS, populations of cells incapable of inducing gp 52 were isolated. When starting with a clone containing multiple proviruses (M1.19), the defect in induction could be ascribed to a lack of functional receptors. Of particular interest, none of the cellular glucocorticoid-inducible proteins, including the enzymes tyrosine aminotransferase, glutamine synthetase, and two cellular glycoproteins called Belt I and Belt II (IVARIE and O'FARRELL 1978), retained hormonal sensitivity (VANNICE et al. 1983). Thus, it appears that the same receptor mechanism involved in regulating cellular genes is responsible for MMTV induction.

When a similar approach was used with a cell line containing a single MMTV provirus (J2.17), variants were isolated that exhibited either reduced or no induction of cell surface gp 52 (GROVE and RINGOLD 1981). These showed a concomitant reduction in intracellular viral RNA and are, therefore, not structural mutants in gp52. Most of the subclones of these variants retain normal inducibility of the cellular glucocorticoid domain, indicating that the defect is specific to MMTV gene expression. By superinfecting such variants with MMTV, VANNICE et al. (1983) could demonstrate that the defect in these clones is *cis* acting (i.e., associated with the original provirus). Molecular cloning of these altered proviruses may provide insights into the mechanistic aspects of glucocorticoid induction of MMTV RNA.

Another immunological approach has recently been used to isolated additional types of variants. FIRESTONE and YAMAMOTO (1983) treated a clone of HTC cells containing multiple MMTV proviruses (M1.54) with antibody to gp52 in the presence of complement. Only hormone-treated cells were killed by this treatment, and, by repeated exposure to this regimen, they obtained two clones with interesting phenotypes. The first exhibits a defect in a glucocorticoid-inducible function required for processing of the envelope glycoproteins; this appears to be a cellular processing function (FIRESTONE et al. 1982), since viral RNA induction appears to be normal. The second clone exhibits a markedly reduced induction of both gp52 and MMTV RNA, as well as a loss of tyrosine aminotransferase induction; however, this clone retains other glucocorticoid-inducible functions as detected by two-dimensional gel electrophoresis. There is currently no explanation for this latter phenotype, although it is conceivable that this clone harbors two independent defects.

## 6.2 Selection of Variant Lymphoma Cells that Survive in Dexamethasone

As described above, many mouse T cell lines die when exposed to glucocorticoids. The mechanism of this killing response is not understood. A novel approach to this problem has recently been presented by DANIELSEN et al. (to be published). These investigators infected a line of mouse T cells (WEHI) with MMTV and could demonstrate that gp52 was induced by dexamethasone prior to cell death. The strategy for selecting mutants defective in the killing pathway, rather than in receptor, was to first isolate a large number of glucocorticoid-resistant cells and then to select those that were still capable of inducing MMTV proteins. This second step utilized the "panning" procedure described by WYSOCKI and SATO (1978), in which antibodies immobilized on a petri dish serve as an immunoadsorbant for antigen-positive cells. Although detailed analyses of these cells remain to be performed, the use of MMTV permitted the first isolation of this class of glucocorticoid-resistant lymphoma variant.

## 7 Transfections with Molecular Clones of Mouse Mammary Tumor Virus DNA

The results of studies on integration and transcription of MMTV DNA (Sects. 5.1 and 5.2) and the genetic studies described in Sect. 6 provide strong evidence that MMTV encodes its own glucocorticoid regulatory region. The advent of molecular cloning of DNA and the ability to introduce such DNAs into tissue culture cells (i.e., transfection), provide powerful new approaches to the study of gene regulation. Mutations or alterations created within a particular DNA sequence in vitro can thus be analyzed phenotypically to identify regulatory regions. Such experiments have been performed using various clones of MMTV DNA introduced into mouse L TK⁻ cells by cotransfection with the herpes simplex virus thymidine kinase (TK) gene (WIGLER et al. 1977, 1979). The results have been consistent, demonstrating that the cloned MMTV DNA becomes stably integrated in the transfected cells, expressed at the level of RNA, and retains glucocorticoid sensitivity. Several different MMTV genomes have been used, including an endogenous provirus from GR mice (HYNES et al. 1981), an endogenous provirus from A/J mice (DIGGELMANN et al. 1982), and an exogenous GR mouse viral DNA cloned from unintegrated circular DNA (BUETTI and DIGGELMANN 1981). Both of the clones representing endogenous and exogenous GR viruses exhibit similar behavior in transfected mink cells, in which the ability of SV40 DNA to oncogenically transform cells was utilized as the selectable marker (OWEN and DIGGELMANN 1983). These studies corroborate the experiments performed by virus infection, indicating that the MMTV genome does, in fact, encode its own hormone regulatory region.

## 7.1 Mouse Mammary Tumor Virus Long Terminal Repeat Contains a Glucocorticoid Regulatory Sequence

The MMTV genome contains approximately 9 kb of DNA, including the 1.3-kb long terminal repeats (Fig. 2). Based on structural considerations and the mechanisms by which retroviral DNA is synthesized (for reviews see TEMIN 1981 and VARMUS 1982), it seemed likely that viral RNA synthesis initiates within the left LTR. Direct sequence analysis of the MMTV LTR and nuclease mapping experiments indeed localize the 5′ end of MMTV RNA to a site approximately 130 nucleotides from the 3′ end of the left LTR (DONEHOWER et al. 1981; LEE et al. 1981; UCKER et al. 1981; KENNEDY et al. 1982; FASEL et al. 1982; MAJORS and VARMUS, unpublished). Since the glucocorticoid-mediated induction of MMTV production appears to be a direct transcriptional effect, chimeric genes containing an MMTV LTR were constructed to test the possibility that the regulatory region resides near the viral promoter. In most cases, the sequence fused to the LTR encodes a protein whose function can be selected for in appropriate recipient cells. Using calcium-phosphate-mediated gene transfer (GRAHAM and VAN DER EB 1973), a plasmid, containing the left LTR fused to a mouse dihydrofolate reductase (DHFR) cDNA (LEE et al. 1981), was introduced into CHO cells deficient in DHFR. Stable transformants containing integrated plasmid DNA were isolated. Using the folate analog, methotrexate, LEE et al. (1981) demonstrated that the production of DHFR was increased three- to fivefold in dexamethasone-treated cells. Furthermore, the 5′ end of the hormone-inducible RNA mapped to the appropriate site within the MMTV LTR.

Similar fusion plasmids containing the *Escherichia coli* xanthine guanine phosphoribosyl transferase (XGPRT) have been introduced into mouse 3T6 cells, using the dominant selection scheme devised by MULLIGAN and BERG (1980, 1981). Again, the production of an MMTV-XGPRT hybrid RNA was glucocorticoid-inducible; inductions ranged from 5- to 15-fold (CHAPMAN et al. 1983).

The right LTR has also been shown to contain a glucocorticoid regulatory region by HUANG et al. (1981). A hybrid between a 2.5-kb MMTV fragment containing the *env* gene region and the right LTR was fused to the transforming gene region of the Harvey murine sarcoma virus. When introduced into NIH/3T3 cells, morphologically transformed colonies were obtained; a rather striking increase in transformants was obtained when cells were exposed to dexamethasone. Direct measurement of the transforming gene product (p 21) indicated a 5- to 30-fold induction by dexamethasone, and, as in the DHFR case, the 5' terminus of the hybrid RNA maps within the MMTV LTR.

Very similar constructions have been used to produce chimeric vectors containing the herpes simplex virus TK gene fused to the right (GRONER et al. 1982) and left (J.E. Majors and H.E. Varmus, unpublished) LTRs. Transfection into mouse L TK⁻ cells allowed selection of stable transformants that exhibit glucocorticoid-inducible TK RNA and enzyme. The constructions described by GRONER et al. (1982) retain the herpes TK promoter intact; the resulting transformants contain two RNAs, one which initiates within the MMTV LTR and the other within the TK gene. It is of interest that only the MMTV-directed RNA is glucocorticoid-inducible.

In addition to the experiments using chimeric genes, FASEL et al. (1982), as well as Yamamoto and colleagues (YAMAMOTO et al. 1981; PAYVAR et al. 1982), have introduced cloned fragments of MMTV DNA into TK⁻ cells by cotransfection with the herpes TK gene. Quantitation of total MMTV-related RNA in these stable transformants was determined by the "dot blot" procedure of KAFATOS et al. (1979). In most, if not all, cases where an intact LTR was introduced, glucocorticoid-inducible RNA could be detected. No mapping studies have been performed to characterize these RNA products. A surprising finding in the experiments of PAYVAR et al. (1982) is that hormone-inducible RNA could be detected in cells transfected with fragments of the MMTV genome derived solely from the *env* gene region and the *gag-pol* gene region. As will be discussed below, specific binding sites for the glucocorticoid-receptor complex may exist on these fragments. It is tempting to speculate that the MMTV genome may harbor the vestiges of multiple glucocorticoid-regulated promoters; in virus-infected cells, however, only the left LTR appears to subserve promoter function.

## 7.2 Glucocorticoid Induction in Transient Expression Assays

Analysis of LTR chimeras in stable transformants has clearly shown that the MMTV LTR contains sequences that confer glucocorticoid sensitivity on downstream genes. However, experiments in which stable transformants must be generated are extremely time consuming. To circumvent this problem, new methods, in which glucocorticoid-inducibility of transfected MMTV sequences can be assayed in a few days, have been developed. HALL et al. (1983) have constructed a series of plasmids in which the *E. coli* β-galactosidase gene is under the control of either the SV40 or MMTV promoter; the assay for this enzyme utilizes a simple colorimetric procedure that is both very sensitive and accurate (MILLER 1972). Using DEAE-dextran to infect mouse L cells with the MMTV-β-gal DNA, they could show that in a short-term, or so-called transient expression, assay, dexamethasone induced the production of β-gal activity at least 20-fold. Enzyme activity was also detected after infection with the SV40-β-gal plasmid; however, it was not inducible by dexamethasone. Since these experiments take only 3–4 days to complete, it should now be possible to determine the effects of mutations and deletions within the LTR rather quickly.

There is evidence that most, if not all, of the DNA found in cells during the first few days after transfection is not integrated (Frankel and Ringold, unpublished). If this is indeed true, it would suggest that integration of MMTV DNA into chromosomal sequences may not be required for glucocorticoid induction. Recent experiments suggest that, in fact, the vast majority of the intracellular plasmid sequences during a transient expression assay exist in the form of autonomous minichromosomes (F.R. Frankel and G.M. Ringold, unpublished). Additional evidence is provided by studies of G. Hager (unpublished) in which glucocorticoid induction from the LTR could be detected from a bovine papilloma virus (BPV) vector that remains in an episomal form even in stable transformants (SARVER et al. 1981). Thus, it seems likely that the chromatin structure of the resident cell is not required for MMTV promoter function or glucocorticoid responsiveness. Characterization of the minichromosomes containing MMTV LTR sequences may reveal interesting information about the function of the glucocorticoid-receptor complex.

## 7.3 Mapping the Glucocorticoid Regulatory Region Within the Long Terminal Repeat

In order to further delineate the sequences of importance for hormonal regulation, several investigators have constructed deletion mutants that remove portions of the LTR in MMTV DNA or in chimeric plasmids. The effects of these alterations have been assessed either in stable transformants or by transient expression assays. As a point of reference, the MMTV LTR contains a so-called Hogness-Goldberg, or TATA, box (GOLDBERG 1978) and a CAAAT sequence, approximately 25 and 65 nucleotides, respectively, upstream from the start of transcription (Fig. 3). These sequences are thought to be important signals for the initiation of transcription (for review, see SHENK 1981). By convention, I will describe the deletions by a negative number corresponding to the residue upstream from the start of MMTV transcription. For example, a molecule that has had all LTR sequences from the left up to 100 nucleotides upstream from the start of transcription removed will be designated as a deletion to residue $-100$.

N.E. HYNES and B. GRONER (in preparation) have constructed a deletion in the left LTR of an intact provirus that extends to approximately residue $-700$. In mouse L cells, both 35- and 24-$S$ RNAs are detected and are glucocorticoid-inducible. In an LTR-herpes TK chimera, they have also constructed a deletion to residue $-270$ that retains hormonal inducibility. Similar TK constructions have been analyzed by J. MAJORS and H. VARMUS (in preparation). They have found that a deletion to residue $-190$ retains inducibility, whereas one to residue $-140$ does not. F. LEE, C. HALL, E. JACOB, D. DOBSON, and G. RINGOLD (in preparation), using a transient $\beta$-gal expression assay, have found that a deletion to $-220$ retains inducibility, whereas a deletion to $-175$ does not. They have also constructed a series of internal deletions that remove nucleotides between residues $-105$ and $-210$. The two most instructive of these delete residues $-105$ to $-140$ and $-105$ to $-210$; the former retains glucocorticoid inducibility, whereas the latter does not. In the aggregate, these data indicate that the left-hand boundary of the glucocorticoid regulatory region must reside between residues $-175$ and $-190$, and that the right-hand boundary must not extend beyond residue $-140$. This is summarized in Fig. 3.

An important point in the analyses of these deletions has been to show that the glucocorticoid regulatory region is distinct and dissociable from the promoter region. Molecules that delete to residues $-140$ or $-109$ retain basal, but not inducible, levels of RNA expression; moreover, the RNA that is made initiates at the known MMTV cap site (J. MAJORS and H. VARMUS, in preparation; F. LEE, C. HALL, E. JACOB, D. DOBSON, and G. RINGOLD, in preparation). A further, perhaps more subtle, point that can be deduced from the analysis of internal deletions is that the absolute spacing between the promoter and regulatory region need not be constant. An insertion of four nucleotides at position $-109$ also does not affect inducibility (Lee, in preparation). It will be of interest to determine what constraints do exist on both the relative spacing between the promoter and regulatory regions, and on the type of sequence (e.g., with respect to secondary structure) that can be placed between the two without affecting hormonal responsiveness.

# 8 Glucocorticoid Receptor and Specific Binding to Mouse Mammary Tumor Virus DNA

## 8.1 Structure of the Glucocorticoid Receptor

Early studies of the glucocorticoid receptor depended on the ability to label the receptor with radioactive steroids. In the past few years, two different approaches have been used to purify the glucocorticoid receptor from rat liver. GOVINDAN and SEKERIS (1978) used an affinity chromatographic procedure, whereas WRANGE et al. (1979) utilized DNA-cellulose chromatography. This latter approach takes advantage of the fact that the receptor will not bind to DNA until activated (i.e., by warming in the presence of hormone). Elution of the glucocorticoid-receptor complex from DNA is facilitated by pyridoxal phosphate, since it appears that the interaction with DNA involves a Schiff's base (CAKE et al. 1978).

Characterization of the purified receptor indicates that it is composed of a single polypeptide chain of approximately 90000 daltons, with an isoelectric point of 5.8 (WRANGE et al. 1979), and is a rather oblong-shaped molecule with a Stokes radius of about 6 nm. Polyclonal antibodies to the receptor have been prepared (GOVINDAN and SEKERIS 1978; OKRET et al. 1981; EISEN 1980), as have monoclonal antibodies (GRANDICS et al. 1982). Their use in characterization of the receptor (CARLSTEDT-DUKE et al. 1982) supports a model proposed earlier for the structural domains of the glucocorticoid receptor (WRANGE and GUSTAFSSON 1978). Briefly, the receptor appears to contain one domain (approximately 40000 daltons) that contains both the steroid- and DNA-binding sites; the remainder of the receptor contains the major antigenic determinants. Based on proteolytic digestion studies, a smaller fragment containing the steroid-binding site can be separated from the DNA-binding region. Analyses of glucocorticoid-receptor mutants of mouse lymphoma cells are in agreement with this model (GEHRING and TOMKINS 1974; YAMAMOTO et al. 1974) and suggest that the approximately 50000-dalton antigenic component of the receptor plays an important role in modulating DNA binding (STEVENS et al. 1981; DELLWEG et al. 1982). Since these experiments deal only with the general DNA binding properties of the glucocorticoid-receptor complex, additional studies will be required to elucidate the role of this region in binding to specific DNA sequences (see below). It is conceivable that this modulatory region plays an important role in stimulating transcription.

## 8.2  Binding of Steroid-Receptor Complexes to Specific DNA Sequences

Several techniques have been used to study the interactions of steroid receptors with specific DNA sequences. GRONEMEYER and PONGS (1980) showed specific binding of ecdysone to puffs on *Drosophila* polytene chromosomes by photoactivated crosslinking of the hormone, presumably in association with its receptor, to the chromatin. The presence of ecdysone at known hormone-responsive puffs was visualized using fluorescent antibodies against the hormone. Using a competition assay, MULVIHILL et al. (1982) have shown specific binding of crude chick progesterone receptors to ovalbumin DNA. Multiple sites on the DNA appear to have high affinity for the receptor. Similar results have been obtained by COMPTON et al. (1982). The relationship between the observed binding and hormonal induction in these systems is not at all clear.


## 8.3 Binding of Glucocorticoid-Receptor Complexes to Mouse Mammary Tumor Virus DNA

Using hormone-receptor complexes purified by repeated DNA cellulose chromatography (WRANGE et al. 1979), PAYVAR et al. (1981, 1982) have shown specific binding of such complexes to MMTV DNA. They have utilized both a nitrocellulose filter binding (RIGGS et al. 1970) and direct visualization in the electron microscope to define four binding regions within cloned MMTV DNA. These are the two LTRs, one region in *pol,* and one region in *env.* Within the LTR, there appear to be four clustered binding sites that reside approximately between residues − 100 and − 400. Similar results have recently been obtained by GEISSE et al. (1982). Based on the functional deletion studies described earlier (Sect. 7.3), it seems likely that only one of these binding sites is required for glucocorticoid induction. The binding regions within *pol* and *env* are somewhat surprising; however, transfection studies (PAYVAR et al. 1982) indicate that hormone-regulated promoters may reside within these fragments. To date, it has not been possible to show that the specific binding observed with this preparation of receptor is hormone dependent.

GOVINDAN et al. (1982) have used affinity-purified receptor to demonstrate specific binding to the LTR. Again, both nitrocellulose filter binding and electron microscopy were used to detect the interaction. With this preparation of receptor, no binding was detected within *pol* or *env,* and a single region residing between − 100 and − 200 in the LTR seemed to contain the major binding site. These investigators report that the binding they are detecting is dependent on the presence of hormone. The discrepancies in the results obtained by these two groups may be attributable to the very different preparations of receptors that were used.

PFAHL (1982) has used a DNA-cellulose competition assay to study the interaction of the glucocorticoid receptor with MMTV DNA. The principle is to compete the binding of crude hormone-receptor complexes to DNA-cellulose with cloned fragments of MMTV DNA. The results indicate that there is a major high-affinity region within the MMTV LTR. No additional sites with comparable affinity could be detected with other fragments of MMTV DNA; a lower-affinity site within the *env* region may be detectable in this assay. The major LTR binding region could be mapped to a fragment containing residues − 50 to − 400.

In sum, although there are some discrepancies that need to be resolved, there is general agreement that one or more high-affinity binding sites for the glucocoticoid-receptor complex are present within the LTR. Moreover, the available data suggest that at least one site maps to the region that is required for glucocorticoid responsiveness in vivo. As might have been anticipated, it seems quite clear that a high-affinity interaction between the receptor and the LTR is involved in the stimulation of MMTV transcription. The detailed mechanisms by which receptor binding elicits this response remain to be elucidated. In light of the fact that DNA in eukaryotic cells is covered with histones and other chromosomal proteins, it will probably be important to study the interaction of the glucocorticoid-receptor complex with the MMTV LTR in its native chromatin configuration.

# 9  Considerations in Future Studies of Transcriptional Regulation of Mouse Mammary Tumor Virus by Glucocorticoids

There has been rapid progress in the study of MMTV transcription and its regulation by glucocorticoids in the past 3 years. The site of transcription initiation has been mapped and the viral RNAs produced have been well characterized. The ability to study wild-type and altered DNA molecules by transfection assays has permitted the identification and rather precise mapping of a glucocorticoid regulatory region within the MMTV LTR. The availability of purified glucocorticoid receptor has allowed direct demonstration that high-affinity binding sites for the hormone-receptor complex exist on MMTV DNA. Isolation of mutant cell lines exhibiting altered hormone-regulated expression of MMTV proteins has provided insights into both cellular and viral processes involved in glucocorticoid action.

Despite these rather prodigious advances, the central question surrounding hormonal regulation of MMTV transcription remains. How does the interaction of the receptor protein with the glucocorticoid regulatory region stimulate transcription from the MMTV promoter? Although specific models must, for the moment, remain sketchy in detail, it is worthwhile considering some possibilities as a focus for future experimentation. The first, and perhaps most obvious, possibility is that direct interactions occur between the receptor and the transcription complex; such protein-protein interactions would facilitate initiation from an appropriately positioned promoter. There is a suggestion that such a mechanism may be involved in positive regulation of transcription by the $\lambda$ phage repressor (M. PTASHNE, personal communication); however, at present there is no compelling reason to accept or reject such a model for steroid action. The fact that there does not appear to be a strict requirement in the spacing between the regulatory region and the MMTV promoter (see Sect. 7.3) might suggest that receptor-polymerase interactions are not of primary importance.

If indeed protein-protein interactions at the promoter region are not involved, then several alternative models for transcriptional regulation might be entertained. The receptor may have the ability to destabilize the DNA double helix near the promoter, thereby allowing more efficient interaction of the RNA polymerase complex with its binding site, or perhaps increasing the probability of forming a stable initiation complex. Such models have been widely discussed in the context of prokaryotic regulatory systems (see McKay and Steitz 1981, for discussion) and have been proposed for the progesterone stimulation of ovalbumin gene transcription (Hughes et al. 1981).

In contrast to direct actions near or at the promoter per se, interaction of the receptor with the regulatory region may affect chromatin structure in a fashion (albeit ill-defined) that permits more efficient initiation of transcription. In this regard, it is noteworthy that alterations in the superhelical density of DNA, apparently imparted by changes in chromatin configuration, seem to play an important role in regulating transcription of the yeast-mating-type locus (Nasmyth et al. 1981; Nasmyth 1982). Moreover, changes in superhelical density have profound effects on sequences that have the potential for existing in alternative secondary structures (Singleton and Wells 1982). It is particularly intriguing that in such a model the action of the receptor at a regulatory site could be envisioned to affect DNA or chromatin structure over long distances.

An alternative model is that the glucocorticoid regulatory region, in association with receptor, serves as an efficient site of entry on the DNA for one or more transcription factors; this factor(s) would be free to scan the DNA, perhaps bidirectionally, for an appropriate promoter site. This model would ascribe to the hormonal regulatory sequences properties akin to the recently described "enhancer" sequences of SV40 and other viral genomes (Banerji et al. 1981; Fromm and Berg 1982; Moreau et al. 1981); these sequences have the amazing ability to exert their effect at distances of several thousand base pairs from the promoter. Recent evidence indicates that enhancer sequences function in a cell- and species-specific manner (Laimins et al. 1982), strongly suggesting that specific proteins must interact with these sequences. In this context, the glucocorticoid regulatory region, when associated with a hormone-receptor complex, may serve as a steroid-dependent enhancer sequence.

Experimental approaches are available to help in distinguishing among these alternatives. It should be possible, for example, to ascertain the minimum and maximum distances by which the regulatory and promoter regions of the LTR can be separated, while maintaining hormone responsiveness. The ability to prepare minichromosomes containing the LTR may allow one to test whether changes in chromatin structure are associated with receptor binding. If the regulatory region behaves as an enhancer, then sequences having promoter activity could block the propagation of the polymerase to the MMTV promoter. Such an observation has recently been made with the SV40 enhancer (J. Banerji and W. Schaffner, unpublished). Whatever the detailed mechanisms involved in glucocorticoid regulation of gene expression may be, it is clear that the MMTV system will continue to play an active role in deciphering the answers. One anticipates that the next 3 years will prove to be as exciting and elucidating as the past 3 have been.

*Acknowledgements.* I thank B. GRONER, N. HYNES, H. DIGGELMANN, K. YAMAMOTO, J. MAJORS, H. VARMUS, C. HALL, and F. LEE for providing results prior to publication. Many fruitful discussions with my colleagues have been invaluable and I thank H. SCHULMAN for his comments on the manuscript. The preparation of this manuscript by F. LINDSAY-FINK is gratefully appreciated.

# References

Ashburner M, Chihara C, Meltzer P, Richards G (1973) Temporal control of puffing activity in polytene chromosomes. Cold Spring Harbor Symp Quant Biol 38:655–662

Banerji J, Rusconi S, Schaffner W (1981) Expression of a beta globin gene is enhanced by remote SV40 DNA sequences. Cell 27:299–308

Baxter JD, Harris AW, Tomkins GM, Cohen M (1971) Glucocorticoid receptors in lymphoma cells in culture: relationship to glucocorticoid killing activity. Science 171:189–191

Bern H, Nandi S (1961) Recent studies of the hormonal influence in mouse mammary tumorigenesis. Prog Exp Tumor Res 2:90–144

Bourne HR, Coffino P, Tomkins GM (1975) Selection of a variant lymphoma cell deficient in adenylate cyclase. Science 187:750–752

Buetti E, Diggelmann H (1981) Cloned mouse mammary tumor virus DNA is biologically active in transfected mouse cells and its expression is stimulated by glucocorticoid hormones. Cell 23:335–345

Cake M, DiSorbo D, Litwack G (1978) Effect of pyridoxal phosphate on DNA binding site of activated hepatic glucocorticoid receptor. J Biol Chem 253:4886–4891

Cardiff RD, Young LJT, Ashley RL (1976) Hormone synergism in the in vitro production of the mouse mammary tumor virus. J Toxicol Environ Health [Suppl]:117–129

Carlstedt-Duke, J, Okret S, Wrange O, Gustafsson J-A (1982) Immunochemical analysis of the glucocorticoid receptor: identification of a third domain separate from the steroid-binding and DNA-binding domains. Proc Natl Acad Sci USA 79:4260–4264

Chapman AB, Costello MA, Lee F, Ringold GM (1983) Amplification and hormone regulated expression of a MMTV-Ecogpt fusion plasmid in mouse 3T6 cells Molec Cell Biol 3:1421–1429

Cohen JC, Shank PR, Morris VL, Cardiff RD, Varmus HE (1979) Integration of the DNA of mouse mammary tumor virus in virus-infected normal and neoplastic tissue of the mouse. Cell 16:333–345

Compton JG, Schrader WT, O'Malley BW (1982) Selective binding of chicken progesterone receptor A subunit to a DNA fragment containing ovalbumin gene sequences. Biochem Biophys Res Commun 105:96–104

Danielsen M, Peterson DO, Stallcup M (to be published) Immunological selection of variant mouse lymphoid cells with altered glucocorticoid responsiveness

Dellweg H-G, Hotz A, Mugele K, Gehring U (1982) Active domains in wild-type and mutant glucocorticoid receptors. EMBO J 1:285–289

Dickson C, Peters G (1981) Protein-coding potential of mouse mammary tumor virus genome RNA as examined by in vitro translation. J Virol 37:36–47

Dickson C, Haslam S, Nandi S (1974) Conditions for optimal MTV synthesis in vitro and the effect of steroid hormones on virus production. Virology 62:242–252

Dickson C, Smith R, Peters G (1981) In vitro synthesis of polypeptides encoded by the long terminal repeat region of mouse mammary tumor virus DNA. Nature 291:511–513

Diggelmann H, Vessaz AL, Buetti E (1982) Cloned endogenous mouse mammary tumor virus DNA is biologically active in transfected mouse cells and its expression is stimulated by glucocorticoid hormones. Virology 122:332–341

Donehower LA, Huang AL, Hager GL (1981) Regulatory and coding potential of the mouse mammary tumor virus long terminal redundancy. J Virol 37:226–238

Dudley JP, Varmus HE (1981) Purification and translation of murine mammary tumor virus mRNAs. J Virol 39:207–218

Duesberg PG, Cardiff RD (1968) Structural relationships between the RNA of mammary tumor virus and those of other RNA tumor viruses. Virology 36:696–700

Eisen HJ (1980) An antiserum to the rat liver glucocorticoid receptor. Proc Natl Acad Sci USA 77:3893–3897

Fasel N, Pearson K, Buetti E, Diggelmann H (1982) The region of mouse mammary tumor virus DNA containing the long terminal repeat includes a long coding sequence and signals for hormonally regulated transcription. EMBO J 1:3–7

Feinstein SC, Ross SR, Yamamoto KR (1982) Chromosomal position effects determine transcriptional potential of integrated mammary tumor virus DNA. J Mol Biol 156:549–565

Fine DL, Plowman JK, Kelly SP, Arthur LO, Hillman EA (1974) Enhanced production of mouse mammary tumor virus in dexamethasone-treated 5-iododeoxyuridine-stimulated mammary tumor cell cultures. JNCI 52:1881–1886

Firestone GL, Yamamoto KR (1983) Two classes of mutant mammary tumor virus-infected HTC cell with defects in glucocorticoid-regulated gene expression. Molec Cell Biol 3:149–160

Firestone GL, Payvar F, Yamamoto K (1982) Glucocorticoid regulation of protein processing and compartmentalization. Nature 300:221–225

Fromm M, Berg P (1982) Deletion mapping of DNA regions required for SV40 early region promoter function in vivo. J Mol Appl Gen 1:457–481

Gehring U, Tomkins GM (1974) A new mechanism for steroid unresponsiveness: loss of nuclear binding activity of a steroid hormone receptor. Cell 3:301–306

Geisse S, Scheidereit C, Westphal HM, Hynes NE, Groner B, Beato M (1982) Glucocorticoid receptors recognize DNA sequences in and around murine mammary tumor virus DNA. EMBO J 1:1613–1619

Gilboa E, Mitra SW, Goff S, Baltimore D (1979) A detailed model of reverse transcription and tests of crucial aspects. Cell 18:93–100

Goldberg ML (1978) Sequence analysis of *Drosophila* histone genes. PhD thesis, Stanford University

Gorski J, Gannon F (1976) Current models of steroid hormone action: a critique. Annu Rev Physiol 38:425–450

Gorski J, Toft DO, Shyamala G, Smith D, Notides A (1968) Hormone receptors: studies on the interaction of estrogen with the uterus. Recent Prog Horm Res 24:45–80

Govindan M, Sekeris C (1978) Purification of two dexamethasone-binding proteins from rat liver cytosol. Eur J Biochem 89:95–104

Govindan MV, Spiess E, Majors J (1982) Purified glucocorticoid receptor hormone complex from rat liver cytosol binds specifically to cloned mouse mammary tumor virus long terminal repeats in vitro. Proc Natl Acad Sci USA 79:5157–5161

Graham FL, van der Eb AJ (1973) A new technique for the assay of infectivity of human adenovirus 5 DNA. Virology 52:456–467

Grandics P, Gasser DL, Litwack G (1982) Monoclonal antibodies to the glucocorticoid receptor. Endocrinol 111:1731–1733

Gronemeyer H, Pongs O (1980) Localization of ecdysterone on polytene chromosomes of *Drosophila melanogaster*. Proc Natl Acad Sci USA 77:2108–2112

Groner B, Hynes NE (1980) Number and location of mouse mammary tumor virus proviral DNA in mouse DNA of normal tissue and of mammary tumors. J Virol 33:1013–1025

Groner B, Hynes NE, Diggelmann H (1979) Identification of MMTV-specific mRNA. J Virol 30:417–420

Groner B, Kennedy N, Rahmsdorf U, Herrlich P, van Ooyen A, Hynes NE (1982) Introduction of a proviral mouse mammary tumor virus gene and a chimeric MMTV-thymidine kinase gene into L cells results in their glucocorticoid responsive expression. In: Dumont JE, Nunez J, Schultz G (eds) Hormones and cell regulation, vol 6. Elsevier, Amsterdam, pp 217–228

Grove JR, Ringold GM (1981) Selection of rat hepatoma cells defective in hormone-regulated production of mouse mammary tumor virus RNA. Proc Natl Acad Sci USA 78:4349–4353

Grove JR, Dieckmann BS, Schroer TA, Ringold GM (1980) Isolation of glucocorticoid-unresponsive rat hepatoma cells by fluorescence-activated cell sorting. Cell 21:47–56

Hall CV, Jacob PE, Ringold GM, Lee F (1983) Expression and regulation of E coli lac Z gene fusions in mammalian cells. J Mol Appl Gen 2:101–109

Higgins SJ, Gehring U (1978) Molecular mechanisms of steroid hormone action. Adv Cancer Res 28:313–397

Huang AL, Ostrowski MC, Berard D, Hager GL (1981) Glucocorticoid regulation of the HaMuSV p21 gene conferred by sequences from mouse mammary tumor virus. Cell 27:245–255

Hughes MR, Compton JG, Schrader WT, O'Malley BW (1981) Interaction of the chick oviduct progesterone receptor with deoxyribonucleic acid. Biochem J 20:2481–2491

Hynes NE, Kennedy N, Rahmsdorf U, Groner B (1981) Hormone-responsive expression of an endogenous proviral gene of mouse mammary tumor virus after molecular cloning and gene transfer into cultured cells. Proc Natl Acad Sci USA 78:2038–2042

Ivarie RD, O'Farrell PH (1978) The glucocorticoid domain: steroid-mediated changes in the rate of synthesis of rat hepatoma proteins. Cell 13:41–55

Jensen EV, Suzuki T, Kawashima T, Stumpf WE, Jungblut PW, DeSombre ER (1968) A two-step mechanism for the interaction of estradiol with rat uterus. Proc Natl Acad Sci USA 59:632–638

Kafatos FC, Jones CW, Efstratiadis A (1979) Determination of nucleic acid sequence homologies and relative concentrations by a dot hybridization procedure. Nucleic Acids Res 7:1541–1552

Kennedy N, Knedlitschek G, Groner B, Hynes NE, Herrlich P, Michalides R, van Ooyen AJJ (1982) The long terminal repeats of an endogenous mouse mammary tumor virus are identical and contain a long open reading frame extending into adjacent sequences. Nature 295:622–624

Klemenz R, Reinhardt M, Diggelmann H (1981) Sequence determination of the 3' end of mouse mammary tumor virus RNA. Mol Biol Rep 7:123–126

Kurtz DT (1981) Hormonal inducibility of rat $\alpha$-2$\mu$ globulin genes in transfected mouse cells. Nature 291:629–631

Laimins LA, Khoury G, Gorman C, Howard B, Gruss P (1982) Host-specific activation of transcription by tandem repeats from simian virus 40 and Moloney murine sarcoma virus. Proc Natl Acad Sci USA 79:6453–6457

Lee F, Mulligan R, Berg P, Ringold G (1981) Glucocorticoids regulate expression of dihydrofolate reductase cDNA in mouse mammary tumor virus chimeric plasmids. Nature 294:228–232

Majors JE, Varmus HE (1981) Nucleotide sequences at host-proviral junctions for mouse mammary tumour virus. Nature 289:253–258

McGrath CM (1971) Replication of mammary tumor virus in tumor cell cultures: dependence on hormone-induced cellular organization. JNCI 47:455–467

McKay DB, Steitz TA (1981) Structure of catabolite gene activator protein at 2.9 Å resolution suggests binding to left-handed B DNA. Nature 290:744–749

McKnight G (1978) The induction of ovalbumin and conalbumin mRNA by estrogen and progesterone in chick oviduct explant cultures. Cell 14:403–413

Michalides R, Wagenaar E, Groner B, Hynes NE (1981) Mammary tumor virus proviral DNA in normal murine tissue and nonvirally induced mammary tumors. J Virol 39:367–376

Miller JH (1972) Experiments in molecular genetics. Cold Spring Harbor Laboratory, Cold Spring Harbor

Moreau P, Hen R, Wasylyk B, Everrett R, Gaub MP, Chambon P (1981) The SV40 72-base-pair repeat has a striking effect on gene expression both in SV40 and other chimeric recombinants. Nucleic Acids Res 9:6047–6069

Mulligan RC, Berg P (1980) Expression of a bacterial gene in mammalian cells. Science 209:1422–1427

Mulligan RC, Berg P (1981) Selection for animal cells that express the Escherichia coli gene coding for xanthine-guanine phosphoribosyltransferase. Proc Natl Acad Sci USA 78:2072–2076

Mulvihill ER, LePennec J-P, Chambon P (1982) Chicken oviduct progesterone receptor: location of specific regions of high-affinity binding in cloned DNA fragments of hormone-responsive genes. Cell 24:621–632

Nandi S, McGrath CM (1973) Mammary neoplasia in mice. Adv Cancer Res 17:353–414

Nasmyth KA (1982) The regulation of yeast-mating-type chromatin structure by SIR: an action at a distance affecting both transcription and transposition. Cell 30:567–578

Nasmyth KA, Tatchell K, Hall BD, Astell C, Smith M (1981) A position effect in the control of transcription at yeast mating-type loci. Nature 289:244–250

Okret S, Carlstedt-Duke J, Wrange O, Carlstrom K, Gustafsson J-A (1981) Characterization of an antiserum against the glucocorticoid receptor. Biochim Biophys Acta 677:205–219

O'Malley R, Roop D, Lai E, Nordstrom J, Catterall J, Swaneck D, Colbert D, Tsai MJ, Dugaiczyk A, Woo S (1979) The ovalbumin gene: organization, structure, transcription, and regulation. Recent Prog Horm Res 35:1–46

Owen D, Diggelmann H (1983) Cloned mouse mammary tumor virus DNA exhibits glucocorticoid-dependent expression in simian virus 40-transformed mink cells. J Viro 45:148–154

Parks WP, Scolnick EM, Kozikowski EH (1974) Dexamethasone stimulation of murine mammary tumor virus expression. A tissue culture source of virus. Science 12:158–160

Payvar F, Wrange O, Carlstedt-Duke J, Okret S, Gustafsson JA, Yamamoto KR (1981) Purified glucocorticoid receptors bind selectively in vitro to a cloned DNA fragment whose transcription is regulated by glucocorticoids in vivo. Proc Natl Acad Sci USA 78:6628–6632

Payvar F, Firestone G, Ross S, Chandler V, Wrange O, Carlstedt-Duke J, Gustafsson J-A, Yamamoto K (1982) Multiple specific binding sites for purified glucocorticoid receptors on mammary tumor virus DNA. J Cell Biochem 19:241–247

Peters G, Glover C (1980) tRNAs and priming of RNA-directed DNA synthesis in mouse mammary tumor virus. J Virol 35:31–40

Pfahl M (1982) Specific binding of the glucocorticoid-receptor complex to the mouse mammary tumor proviral promoter region. Cell 31:475–482

Pfahl M, Kelleher RJ, Bourgeois S (1978) General features of steroid resistance in lymphoid cell lines. Mol Cell Endocrinol 10:193–207

Proudfoot NJ, Brownlee GG (1974) Sequence at the 3' end of globin mRNA shows homology with immunoglobulin light chain mRNA. Nature 252:359–362

Riggs AD, Suzuki H, Bourgeois S (1970) lac repressor-operator intraction I. Equilibrium studies. J Mol Biol 48:67–83

Ringold GM, Lasfargues EY, Bishop JM, Varmus HE (1975a) Production of mouse mammary tumor virus by cultured cells in the absence and presence of hormones: assay by molecular hybridization. Virology 65:135–147

Ringold GM, Yamamoto KR, Tomkins GM, Bishop JM, Varmus HE (1975b) Dexamethasone-mediated induction of mouse mammary tumor virus RNA: a system for studying glucocorticoid action. Cell 6:299–305

Ringold GM, Cardiff RD, Varmus HE, Yamamoto KR (1977a) Infection of cultured hepatoma cells by mouse mammary tumor virus. Cell 10:11–18

Ringold GM, Yamamoto KR, Bishop JM, Varmus HE (1977b) Glucocorticoid-stimulated accumulation of mouse mammary tumor virus RNA: increased rate of synthesis of viral RNA. Proc Natl Acad Sci USA 74:2879–2883

Ringold G, Yamamoto K, Shank P, Varmus H (1977c) Mouse mammary tumor virus DNA in infected cells: characterization of unintegrated forms. Cell 10:19–26

Ringold GM, Shank PR, Yamamoto KR (1978) Production of unintegrated mouse mammary tumor virus DNA in infected rat hepatoma cells is a secondary action of dexamethasone. J Virol 26:93–101

Ringold GM, Shank PR, Varmus HE, Ring J, Yamamoto KR (1979) Integration and transcription of mouse mammary tumor virus DNA in rat hepatoma cells. Proc Natl Acad Sci USA 76:665–669

Robertson DL, Varmus HE (1979) Structure and function of the intracellular RNAs of murine mammary tumor virus. J Virol 30:576–589

Rosenau W, Baxter J, Rousseau G, Tomkins G (1972) Mechanism of resistance to steroids: glucocorticoid receptor defect in lymphoma cells. Nature 237:20–22

Ross E, Gilman A (1977) Resolution of some components of adenylate cyclase necessary for catalytic activity. J Biol Chem 252:6966–6969

Sarver N, Gruss P, Law M-F, Khoury G, Howley P (1981) Bovine papilloma virus deoxyribo-
    nucleic acid: a novel eukaryotic cloning vector. Mol Cell Biol 1:486–496
Schimke R, McKnight GS, Shapiro D, Sullivan D, Palacios R (1975) Hormonal regulation
    of ovalbumin synthesis in the chick oviduct. Recent Prog Horm Res 31:175–209
Scolnick EM, Young HA, Parks WP (1976) Biochemical and physiological mechanisms in
    glucocorticoid hormone induction of mouse mammary tumor virus. Virology 69:148–156
Sen GC, Smith SW, Marcus SL, Sarkar NH (1979) Identification of the messenger RNAs
    coding for the gag and env gene products of the MTV. Proc Natl Acad Sci USA
    76:1736–1740
Shank PR, Cohen JC, Varmus HE, Yamamoto KR, Ringold GM (1978) Mapping of linear
    and circular forms of mouse mammary tumor virus DNA with restriction endonucleases:
    evidence for a large specific deletion occurring at high frequency during circularization.
    Proc Natl Acad Sci USA 75:2112–2116
Shenk T (1981) Transcription control regions: nucleotide sequence requirements for initiation
    by RNA polymerase II and III. Curr Top Microbiol Immunol 93:25–40
Shyamala G (1973) Specific cytoplasmic glucocorticoid hormone receptors in lactating mamma-
    ry glands. Biochemistry 12:3085–3090
Shyamala G, Dickson C (1976) Relationship between receptor and mammary tumor virus
    production after stimulation by glucocorticoid. Nature 262:107–112
Sibley CH, Tomkins GM (1974a) Isolation of lymphoma cell variants resistant to killing
    by glucocorticoids. Cell 2:213–220
Sibley CH, Tomkins GM (1974b) Mechanisms of steroid resistance. Cell 2:221–227
Singleton CK, Wells RD (1982) Relationship between superhelical density and cruciform for-
    mation in plasmid pVH51. J Biol Chem 257:6292–6295
Smoller CG, Pitelka DR, Bern HA (1961) Cytoplasmic inclusion bodies in cortisol-treated
    mammary tumors of C3H/Crgl mice. J Biophys Biochem Cytol 9:915–920
Southern EM (1975) Detection of specific sequences among DNA fragments separated by
    gel electrophoresis. J Mol Biol 38:503–517
Stallcup MR, Ring J, Yamamoto KR (1978) Synthesis of mouse mammary tumor virus ribonu-
    cleic acid in isolated nuclei from cultured mammary tumor cells. Biochemistry
    17:1515–1521
Stevens J, Eisen HJ, Stevens Y-W, Haubenstock H, Rosenthal RL, Artishevsky A (1981)
    Immunochemical differences between glucocorticoid receptors from corticoid-sensitive and
    resistant malignant lymphomas. Cancer Res 41:134–137
Temin HM (1981) Structure, variation and synthesis of retrovirus long terminal repeat. Cell
    27:1–3
Topper YJ, Freeman CS (1980) Multiple hormone interactions in the developmental biology
    of the mammary gland. Physiol Rev 60:1049–1106
Ucker DS, Ross SR, Yamamoto KR (1981) Mammary tumor virus DNA contains sequences
    required for its hormone-regulated transcription. Cell 27:257–266
Vaidya AB, Lasfargues, EY, Heubel G, Lasfargues JC, Moore DH (1976) Murine mammary
    tumor virus: characterization of infection of non-murine cells. J Virol 18:911–917
Vannice JL, Grove JR, Ringold GM (1983) Analysis of glucocorticoid-inducible genes in
    wild-type and variant rat hepatoma cells. Mol Pharmacol 23:779–785
Varmus HE (1982) Form and function of retroviral proviruses. Science 216:812–820
Weintraub H, Groudine M (1976) Chromosomal subunits in active genes have an altered
    conformation. Science 193:848–856
Wigler M, Silberstein S, Lee LS, Pellicer A, Cheng YC, Axel R (1977) Transfer of purified
    herpesvirus thymidine kinase gene to cultured mouse cells. Cell 11:223–232
Wigler M, Sweet R, Sim GK, Wold B, Pellicer A, Lacy E, Maniatis T, Silberstein S, Axel
    R (1979) Transformation of mammalian cells with genes from prokaryotes and eukaryotes.
    Cell 16:777–785
Wrange O, Gustafsson J-A (1978) Separation of the hormone- and DNA-binding sites of
    the hepatic glucocorticoid receptor by means of proteolysis. J Biol Chem 253:856–865
Wrange O, Carlstedt-Duke J, Gustafsson J-A (1979) Purification of the glucocorticoid receptor
    from rat liver cytosol. J Biol Chem 254:9284–9290
Wysocki LJ, Sato VL (1978) "Panning" for lymphocytes: a method for cell selection. Proc
    Natl Acad Sci USA 75:2844–2848

Yamamoto KR, Alberts BM (1976) Steroid receptors: elements for modulation of eukaryotic transcription. Annu Rev Biochem 45:721–746

Yamamoto KR, Stampfer MR, Tomkins GM (1974) Receptors from glucocorticoid-sensitive lymphoma cells and two classes of insensitive clones: physical and DNA-binding properties. Proc Natl Acad Sci USA 71:3901–3905

Yamamoto KR, Chandler VL, Ross SR, Ucker DS, Ring JC, Feinstein SC (1981) Integration and activity of mammary tumor virus genes: regulation by hormone receptors and chromosomal position. Cold Spring Harbor Symp Quant Biol 45:687–697

Yang J, Enami J, Nandi S (1977) Regulation of mammary tumor virus production by prolactin in Balb/c fC3H mouse normal mammary epithelial cells in vitro. Cancer Res 37:3644–3647

Young HA, Scolnick EM, Parks WP (1975) Glucocorticoid receptor interactions and induction of murine mammary tumor virus. J Biol Chem 250:3337–3343

Young HA, Shih TY, Scolnick EM, Parks WP (1977) Steroid induction of mouse mammary tumor virus: effect upon synthesis and degradation of viral. RNA. J Virol 21:139–146